# Handbook of
# Practical
# Electronic Tests
# and Measurements

PRENTICE-HALL SERIES IN ELECTRONIC TECHNOLOGY

Dr. Irving L. Kosow, editor

Charles M. Thomson, Joseph J. Gershon and Joseph A. Labok
consulting editors

PRENTICE-HALL INTERNATIONAL, INC., *London*
PRENTICE-HALL OF AUSTRALIA, PTY. LTD., *Sydney*
PRENTICE-HALL OF CANADA, LTD., *Toronto*
PRENTICE-HALL OF INDIA PRIVATE LTD., *New Delhi*
PRENTICE-HALL OF JAPAN, INC., *Tokyo*

# Handbook of Practical Electronic Tests and Measurements

John D. Lenk

*Consulting Technical Writer*

PRENTICE-HALL, INC.
*Englewood Cliffs, N.J.*

Current printing (last digit):

10   9

13-380626-X
Library of Congress Catalog Card Number: 74-82907
Printed in the United States of America

*Dedicated to my wife Irene
and daughter Karen*

# Preface

As the title implies, this is a handbook of *practical* electronic test and measurement procedures. The handbook describes procedures for *test of electronic components* (such as antennas, capacitors, coils, diodes, resistors, transformers, transistors, and so on), as well as procedures for *measurement of quantities* (such as impedance, inductance, frequency response, circuit Q, distortion, standing-wave ratio, and so on).

The procedures described can be performed with *basic* test equipment available in all electronic laboratories and most shops. Procedures requiring specialized test equipment (radar, microwave, electromechanical) are generally omitted. The reason for such omission is that the instruction manuals supplied with specialized test equipment generally describe both the operation and application in great detail.

Each procedure described in this handbook is preceded by a brief description of the "why" and "where" of the particular test. These descriptions offer a digest for readers who may be unfamiliar with some characteristic of a particular component or quantity and want to put the step-by-step procedures to immediate use. The combination of theory-plus-application makes this handbook suitable as a reference text for student technicians and as a guidebook for experienced, working technicians.

Although every known electronic test procedure has not been included, all practical, experience-proven applications are presented. At least one procedure is included for test of all basic electronic components, as well as for all commonly used quantities. Where practical, two procedures are given for each test or measurement. One procedure is for "quick test," using the simplest of equipment. The other procedure is primarily for laboratory work and requires more elaborate equipment.

It is assumed that the reader is already familiar with operation and use of basic electronic test equipment such as meters, oscilloscopes, and signal generators. Such information is covered in the author's *Handbook of Oscilloscopes,* and *Handbook of Electronic Meters,* published by Prentice-Hall.

The author has received much help from various organizations in writing this handbook. He wishes to give special thanks to the following: Allied Radio, B & K Manufacturing, Cohu Electronics, Inc., Eico, General Electric, Hewlett-Packard Co., International Rectifier Corporation, Motorola, Non-Linear Systems, Inc., Radio Corporation of America, Sencore, Simpson Electric Co., Society of Motion Picture and Television Engineers, Tektronix Inc., Texas Instruments, and Triplett Electrical Instrument Co.

J. D. L.

# Contents

## 10   MICROWAVE MEASUREMENTS   221

## 11   ANTENNA AND TRANSMISSION LINE MEASUREMENTS   237

## 12   MISCELLANEOUS CIRCUIT AND COMPONENT TESTS   255

# Handbook of
# Practical
# Electronic Tests
# and Measurements

# Safety Precautions

Certain precautions must be observed during operation of any electronic test equipment. Many of these precautions are the same for all types of test equipment; other precautions are unique to special test instruments such as meters, oscilloscopes, signal generators, etc. Some of the precautions are designed to prevent damage to the test equipment or the circuit under test; others are to prevent injury to the operator.

The following general safety precautions should be studied thoroughly and then compared to any specific precautions called for in the equipment instruction manuals.

1. Many test instruments are housed in metal cases. These cases are connected to the ground of the internal circuit. For proper operation, the ground terminal of the instrument should always be connected to the ground of the equipment under test. Make certain that the chassis of the equipment under test is not connected to either side of the a-c line (as is the case with some older a-c/d-c radio sets) or to any potential above ground. If there is any doubt, connect the equipment under test to the power line through an *isolation transformer*.

2. Remember, there is always danger inherent in testing electrical equipment that operates at hazardous voltages. Therefore, the operator should familiarize himself thoroughly with the equipment under test before working on it, bearing in mind that high voltage may appear at unexpected points in defective equipment.

3. It is good practice to remove power before connecting test leads to high-voltage points. (High-voltage probes are often provided with alligator clips.) It is preferable to make all test connections with the power removed. If this is impractical, be especially careful to avoid accidental contact with equipment and other objects that can provide a ground. Working with one hand in your pocket and standing on a properly insulated floor lessens the danger of shock.

4. Filter capacitors may store a charge large enough to be hazardous. Therefore, discharge filter capacitors before attaching the test leads.

5. Remember that leads with broken insulation offer the additional hazard of high voltages appearing at *exposed* points along the leads. Check test leads for frayed or broken insulation before working with them.

6. To lessen the danger of accidental shock, disconnect test leads immediately after the test is completed.

7. Remember that the risk of severe shock is only one of the possible hazards. Even a minor shock can place the operator in danger of more serious risks, such as a bad fall or contact with a source of higher voltage.

8. The experienced operator guards continuously against injury and does not work on hazardous circuits unless another person is available to assist in case of accident.

9. Even if you have had considerable experience with test equipment, always study the instruction manual of any instrument with which you are not familiar.

10. Use only shielded probes and test leads. Never allow your fingers to slip down to the metal probe tip when the probe is in contact with a "hot" circuit.

11. Avoid vibration and mechanical shock. Most electronic test equipment is somewhat delicate.

12. Study the circuit under test *before making any test connections*. Try to match the capabilities of the instrument to the circuit under test. For example, if the circuit under test has a range of measurements to be made (ac, dc, RF, modulated signals, pulses or complex waves), it may be necessary to use more than one instrument. As another example, most meters will measure dc and low-frequency ac. If an unmodulated RF carrier is to be measured, use an RF probe. If the carrier to be measured is modulated with low-frequency signals, a demodulator probe must be used. If pulses, square waves, or complex waves (combinations of ac, dc, and pulses) are to be measured, a peak-to-peak reading meter, or an oscilloscope, will provide the only *meaningful* indications.

# CHAPTER 1

# Voltage Measurements

## 1-1. Kirchhoff's Voltage Law

In many electrical circuits, the arrangement of the devices and applied voltages makes it almost impossible to solve such networks by simple application of Ohm's law. These problems can often be solved by applying Kirchhoff's voltage law. This law can be stated in the following manner. *In any complete electrical circuit, the sum of all the voltages and all the voltage (IR) drops, taken with their proper signs, is zero;* or, the sum of the voltage drops in a complete circuit is equal to the applied voltage. Figure 1-1 shows the relationship of the voltages and currents in a typical series circuit.

The following pointers will help in applying Kirchhoff's law to practical voltage measurements.

1. In solving theoretical problems, it is not always easy to determine the direction of current. Therefore, *assume* a current flow for theoretical problems. If the assumption is backwards, the answer for current magnitude will be numerically correct but will be a *negative* number.

2. In practical problems, the direction of current flow will be found automatically (the voltmeter will show a negative or reverse indication if the test leads are not connected properly). However, the voltage across each resistance must be measured separately, and the correct polarity must be observed.

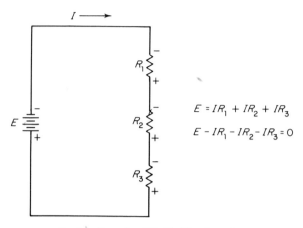

**Fig. 1-1.** Example of Kirchhoff's voltage law.

3. In theory, *place polarity markings* on all voltage sources and resistors in the circuit. The assumed current direction will not affect the voltage source polarities, but the voltage drop on resistors will be affected.

4. In practice, make certain to measure all voltage sources and all voltage drops.

5. In theory or practice, work around the circuit and set up each term of the equation. In a theoretical equation, precede each term by the sign found on *leaving* each particular voltage source or resistor. In practice, use the actual signs found by voltage measurement.

In the example of Figure 1-1 (a series circuit), the same current flows through each resistor. If the voltage (IR) drops across each resistor are added, their sum will be equal to the voltage of the source. If this sum is subtracted from the source voltage, the result will be zero. Also note that the polarity of the battery voltage and the polarity of the voltage (IR) drops across the resistors are opposite.

### 1-1.1. Effect of Aiding and Opposing Voltages in Series

When two or more voltage sources are connected in series, their effect is additive if the polarities are aiding. If the polarities are opposed, the smaller voltage is subtracted from the larger. If there are several voltages of various polarities, the total of one polarity is subtracted from the total of the other polarity. Figure 1-2 shows these relationships in typical circuits.

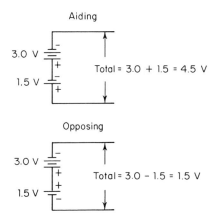

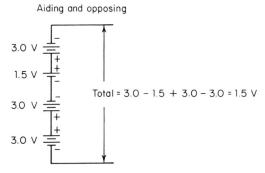

**Fig. 1-2.** Effect of aiding and opposing voltages in series.

## 1-2. Basic Voltmeter Measurements

The following paragraphs describe the *basic* steps necessary to measure voltage with a volt-ohmmeter (VOM) or electronic voltmeter (VTVM, transistorized voltmeter, etc.).

The first step in making a voltage measurement is to set the range. Always use a range that is *higher* than the anticipated voltage. If the approximate voltage is not known, use the highest voltage range initially, then select a lower range so that a good mid-scale reading can be obtained.

Set the meter function selector to ac or dc as required. In the case of dc, it may also be necessary to select either plus or minus by means of the function switch. On simple meters, polarity is changed by switching the test leads.

On an electronic voltmeter, the next step is to zero the meter. This should be done after the range and function have been selected. Touch the test leads together and adjust the "zero" control for a zero indication on the voltage scale to be used.

One common problem in any voltage measurement is that there may be both ac and dc in the circuit being measured. The following summarizes the various aspects of this problem.

If it is desired to measure only ac in the presence of dc, the "output" or "ac only" function can be selected, in which a coupling capacitor is switched into the input circuit. On a VOM, this is done by connecting the free test lead to the "output" terminal. In an electronic voltmeter, ac is often selected by means of a switch on the probe. In some meters, ac is always measured with a coupling capacitor at the input. In any event, the dc is blocked, and the ac is passed.

Use of the output function can present another problem. The coupling capacitor and the meter resistance form a high-pass filter and may attenuate low-frequency a-c voltages. However, most meters will provide accurate a-c indications above 15 or 20 Hz. It is also possible that the coupling capacitor and the meter movement coil will form a resonant circuit and may increase the a-c signals at some particular frequency (usually about 30 to 60 kHz).

Always consider the frequency problem when making any a-c voltage measurements.

If it is desired to measure only dc in the presence of ac, there are several possibilities. If the ac is of high frequency, it is possible that the meter movement will not respond and that no ac indications will occur when the meter is set to measure dc. If the amplitude of the ac is low in relation to the level of the dc being measured, it is also possible that the meter will not be affected.

One solution is to connect a capacitor across the test leads. This will provide a bypass for the ac but will not affect the dc. However, the capacitor may affect operation of the circuit. Remember that the capacitor will be charged to the full value of the dc.

In some cases a high-voltage or attenuator probe may be used to measure dc in the presence of ac. The series resistance of the probe, combined with the natural capacitance between the probe's inner and outer conductor or shield, forms a low-pass filter. This filter action will have no effect on dc but will reject ac.

The fact that electronic voltmeters usually use some form of probe makes these instruments better suited to measure dc (in the presence of ac) than VOMs.

Remember that all voltage measurements (ac, dc, plus, minus, decibels) are made with the meter in *parallel* across the circuit and voltage source, as shown in Fig. 1-3. This means that some of the current normally passing through the circuit under test will be passed through the meter.

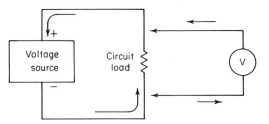

**Fig. 1-3.** Voltage measurements are made with meter in parallel across circuit and voltage source.

In a VOM in which the total meter resistance (or impedance) is low, considerable current may pass through the meter. This may or may not affect the circuit operation. For example, an oscillator that develops a small voltage over a high circuit impedance can be prevented from oscillating if a VOM is used to measure the voltage. (The low-impedance VOM draws excessive current, dropping the voltage to a point where oscillator feedback cannot occur.)

This problem of parallel current drain does not occur in an electronic voltmeter, except where a voltage is measured across a high impedance circuit. A typical electronic voltmeter will have an input impedance of 10 to 15 megohms. If the circuit impedance is near this value, the current will divide itself between the circuit and the meter, resulting in an error.

## 1-3. Checking Voltmeter Accuracy

Both the a-c and d-c scales of a voltmeter must be checked for voltage accuracy. In addition, the a-c scales must be checked for accuracy over the entire rated frequency range. Since this handbook is concerned primarily with test and measurement procedures and assumes that the reader is familiar with test equipment calibration, no detailed voltage calibration procedures are given here. The author's *Handbook of Electronic Meters: Theory and Application* (Englewood Cliffs, N.J.: Prentice-Hall, Inc., 1969) provides a comprehensive discussion of methods for voltage calibration. The following paragraphs describe *basic* procedures for checking voltmeter accuracy.

### 1-3.1. Voltage Accuracy

The obvious method for checking the accuracy of a voltmeter is to measure a known voltage or series of voltages and compare the indicated voltage values with tolerances. There are two basic methods for making such a test.

The most convenient method is to compare the voltmeter to be tested against a standard voltmeter of known accuracy. It is common practice in laboratory work to have one standard voltmeter against which all meters are compared. The standard voltmeter is never used for routine work, but only for testing, and it is sent out for calibration against a *primary standard* at regular intervals.

The voltmeters to be tested are connected in parallel with the standard voltmeter and a variable voltage source, as shown in Fig. 1-4. The source is then varied over the entire range of the voltmeter, and the voltage indications are compared. In the case of an a-c meter, the source is set to a given voltage; then the frequency is varied over the entire range of the meter.

In the type of test shown in Fig. 1-4, accuracy is dependent entirely on the accuracy of the standard meter, not on the source voltage or frequency. This is very convenient, since it is more practical to maintain a meter of known accuracy than to keep a source of known accuracy, especially a source that can be varied over the entire voltage and frequency range of a typical VOM or electronic meter.

The next most popular method of checking voltage accuracy is to measure a voltage source of known accuracy. Inexperienced operators often check the voltmeter accuracy against a common dry cell or series of dry cells. This is satisfactory for rough shop work but is not accurate enough for precision work. A single dry cell rarely produces 1.5 V, as is often assumed. The accuracy of a dry-cell voltage is usually less than that of a typical VOM.

A *mercury cell* or series of mercury cells provides much better accuracy than a common dry cell. The most accurate source is a "standard cell," such as the *Weston cell.* Inexperienced operators will often connect a meter to be tested directly to a standard cell. While this will provide an accurate indication, it will also place a damaging current drain on the standard cell. Most laboratories use some form of *calibration circuit* in combination with the standard cell.

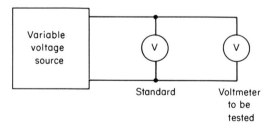

Fig. 1-4. Basic voltmeter calibration circuit.

## 1-4. Extending Voltmeter Ranges

The range of a voltmeter can be extended to measure high voltages by using a *high-voltage probe* or by using *external multiplier resistors* connected to the basic meter movement, as shown in Fig. 1-5. In most meters, the basic movement is used on the *lowest* current range.

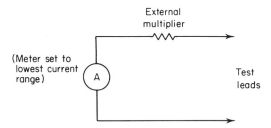

**Fig. 1-5.** Extending meter voltage ranges with an external multiplier.

Values for multipliers can be determined using the procedures given below.

### 1-4.1. Measuring Internal Resistance of Meter Movements

To calculate the values for multipliers it is necessary to know the current required for full-scale deflection and the resistance of the basic meter movement. Usually, the full-scale deflection current is indicated on the scale face. However, the internal resistance is usually not marked. Therefore, the first step in calculating multiplier values is to find the internal resistance of the meter movement by using the test circuit of Fig. 1-6 or 1-7.

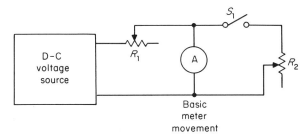

**Fig. 1-6.** Measuring internal resistance of meter movement (half-scale method).

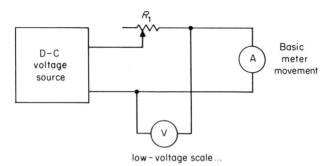

**Fig. 1-7.** Measuring internal resistance of meter movement (voltage-current method).

To use the circuit of Fig. 1-6, disconnect $R_2$ from the circuit by open-ing switch $S_1$. Adjust $R_1$ until the meter movement is at full-scale deflec-tion. Close switch $S_1$, and adjust $R_2$ until the meter movement is at exactly one-half scale deflection. Disconnect $R_2$ from the circuit, and measure the $R_2$ resistance value with an ohmmeter. The $R_2$ resistance value is equal to the internal resistance of the meter movement.

To use the circuit of Fig. 1-7, adjust $R_1$ until the meter movement is at approximately 75% of full-scale deflection. Adjust $R_1$ for some exact current value. Calculate the internal resistance using Ohm's law ($R = E/I$). Note that the voltage drop across a typical movement will be less than one volt. Therefore, it will probably be necessary to use the lowest scale of the voltmeter. *Never connect an ohmmeter directly across the meter movement. This will damage (and probably burn out) the movement.*

### 1-4.2. Determining Multiplier Resistance Values

The multiplier resistance required to convert a basic meter movement into a voltmeter capable of measuring a given voltage range can be cal-culated using the following equation

$$R_x = \frac{R_m (V_2 - V_1)}{V_1}$$

where $R_x$ is the multiplier resistance (in series with the meter movement),

$R_m$ is the internal resistance of the meter movement,

$V_1$ is the voltage required for full-scale deflection of the meter movement, and

$V_2$ is the voltage desired for full-scale deflection (maximum voltage range desired).

For example, assume that a meter has a 50 $\mu$A full-scale movement (with 50 equal divisions on the scale), an internal resistance of 300 ohms, and it is desired to convert the meter movement to measure 0 to 100 V. (Each of the 50 scale divisions will then represent 2 V.)

1. Find the voltage required for full-scale deflection of the meter movement: $E = IR$ or $(5 \times 10^{-5}) (3 \times 10^2) = 0.015$ V. This is voltage $V_1$.
2. Using the equation given above

$$R_x = \frac{300(100 - 0.015)}{0.015} = 1,999,700 \text{ ohms}$$

3. To verify this multiplier resistance and the meter internal resistance (300 ohms), divide by the full-scale voltage obtained with the multiplier (100 V) and find the ohms-per-volt rating:

$$1,999,700 + 300 = \frac{2,000,000}{100} = 20,000 \text{ ohms/V.}$$

4. Then divide the full-scale current (50 $\mu$A) into one and find an ohms-per-volt rating of 20,000 (both ohms-per-volt ratings should match).

### 1-4.3. Measuring Very Low Voltages

The basic meter movement can be used to measure very low voltages. However, *great care* must be used not to exceed the voltage drop required for full-scale deflection of the basic movement. For example, if we assume that a 100-$\mu$A movement has a 2500-ohm internal resistance, the voltage drop required for full-scale deflection is 0.25 V. Voltages greater than this will destroy the meter. To be safe, make a rough check of the voltage to be measured on one of the meter's regular voltage scales. If the indication appears well below the full-scale deflection voltage of the basic meter movement, it is safe to proceed.

## 1-5. Suppressed-Zero Voltage Measurements

Sometimes it is convenient to measure a large voltage source that is subject to small variations on a low voltage range. Thus, small voltage differences can be measured easily. For example, assume that a 100-V source is subject to 1-V variations (99 to 101 V). This would show up as one scale division on a 100-V scale. It is possible to use the suppressed-zero technique and measure the difference on a low scale, such as the 2.5- or 3-V scale found on most meters.

The basic suppressed-zero circuit is shown in Fig. 1-8. Note that an opposing voltage is connected in series with the voltage to be measured. Therefore, the meter sees only the *difference* in voltage. For example, assume that a source that varies from 99 to 101 V is to be measured on

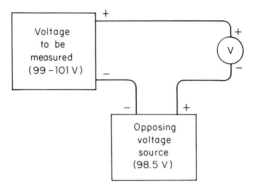

**Fig. 1-8.** Basic suppressed-zero voltage measurement technique.

the 3-V scale of a VOM. A 98.5-V battery arrangement can be used as the opposing source. If the source to be measured is at 99 V, the meter reads 0.5 V (99 — 98.5). If the source is at 101 V, the meter reads 2.5 V (101 — 98.5). Both of these indications can be read easily on the 3-V scale.

Note that the suppressed-zero technique does not increase accuracy of voltage measurements but simply makes small voltage variations easy to read.

## 1-6. Measuring Voltages with Oscilloscopes

The oscilloscope has both advantages and disadvantages when used to measure voltage (and current). The most obvious advantage is that the oscilloscope shows waveform, frequency, and phase simultaneously with the amplitude of the voltage (or current) being measured. The VOM or electronic voltmeter shows only amplitude. Likewise, most meters are calibrated in relation to sine waves. When the signals being measured contain significant harmonics, the calibrations are inaccurate. With the oscilloscope, the voltage is measured from the displayed wave, which includes any harmonic content. In certain applications, the lack of inertia and the high-speed response of an oscilloscope make it the only instrument capable of transient-voltage measurement.

The only major disadvantage of using an oscilloscope for voltage (or current) measurement is the problem of resolution. The scales of simple, inexpensive VOMs or electronic voltmeters are easier to read than the graticules of an oscilloscope. In most cases, the oscilloscope's vertical scales are used for voltage (or current) measurements, with each scale division representing a given value of voltage or current. Where voltages are large, it is difficult to interpolate between divisions.

Another problem, although not a disadvantage, is that voltages measured with an oscilloscope are peak-to-peak, whereas most voltages specified in electronic maintenance and troubleshooting manuals are RMS. This requires that the peak-to-peak value be converted to RMS.

To sum up, if the only value of interest is voltage (or current) amplitude, use the meter because of its simplicity in readout. Use the oscilloscope when waveshape characteristics are of equal importance to amplitude.

### 1-6.1. Voltage Calibration

The vertical amplifier of a laboratory oscilloscope usually has a step-attenuator in which each step is related to a specific deflection factor (such as volts per centimeter). Many such oscilloscopes have a vertical gain adjust that, when adjusted against an internal calibrated source, sets the accuracy of the attenuator. These oscilloscopes need not be calibrated for voltage (or current) measurements, since calibration is an internal adjustment performed as part of routine maintenance.

The vertical amplifiers of shop oscilloscopes usually have variable attenuators and possibly a step-attenuator. The steps do not, however, have a specific volts-per-centimeter deflection factor. Such oscilloscopes must be calibrated before they can be used to measure voltage (and current).

Since this handbook is concerned primarily with test and measurement procedures and assumes that the reader is familiar with test equipment calibration, no detailed voltage calibration procedures are given here. The author's *Handbook of Oscilloscopes: Theory and Application* (Englewood Cliffs, N.J.: Prentice-Hall, Inc., 1968) provides a comprehensive discussion of methods for oscilloscope calibration.

## 1-7. Measuring Peak-to-peak Voltages with an Oscilloscope

The following procedures describe the general steps necessary to measure peak-to-peak voltages with an oscilloscope. Of course, the specific steps will depend upon the operating controls of the particular oscilloscope.

1. Connect the equipment as shown in Fig. 1-9.

2. Place the oscilloscope in operation as described in the instruction manual.

3. On a laboratory oscilloscope, set the vertical step-attenuator to a deflection factor that will allow the expected signal to be displayed without overdriving the vertical amplifier.

4. On a shop oscilloscope, set the vertical-gain control to the "calibrate-set" position as determined during the calibration procedure.

5. Set the input selector (if any) to measure ac. Connect the probe to the signal being measured.

6. Switch on the oscilloscope internal recurrent sweep. Adjust the sweep frequency for several cycles on the screen.

7. Adjust the horizontal control to spread the pattern over as much of the screen as desired.

8. Adjust the vertical position control so that the downward excursion of the waveform coincides with one of the graticule lines below the graticule center line, as shown in Fig. 1-9.

## NOTE

On a shop oscilloscope, do not move the vertical gain control from the "calibrate-set" position. Use the vertical position control only.

9. Adjust the horizontal position control so that one of the upper peaks of the signal lies near the vertical center line, as shown in Fig. 1-9.

10. Measure the peak-to-peak vertical deflection in centimeters.

## NOTE

This technique may also be used to make vertical measurements between two corresponding points on the waveform other than peak-to-peak. The peak-to-peak points are usually easier to measure.

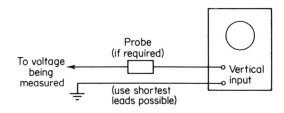

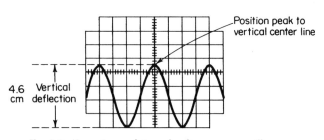

**Fig. 1-9.** Measuring peak-to-peak voltages on an oscilloscope.

11. On a laboratory oscilloscope, multiply the distance measured in Step 10 by the vertical attenuator switch setting. Multiply the result by the attenuation factor of the probe, if any. This gives the peak-to-peak voltage.

12. On a shop oscilloscope, multiply the distance measured in Step 10 by the calibration factor (in volts per centimeter) established during calibration. Multiply the result by the attenuation factor of the probe (if any) and the setting of the attenuator step switch (if any) to get peak-to-peak voltage.

For example, assume a peak-to-peak vertical deflection of 4.6 cm (Figure 1-9), using a $10\times$ attenuator probe, and a vertical-deflection factor of 0.5 V/cm.

We use the general equation

$$\text{volts peak-to-peak} = \begin{matrix}\text{vertical-}\\\text{deflection}\\\text{factor}\end{matrix} \times \begin{matrix}\text{volts-per-centimeter}\\\text{factor}\end{matrix} \times \begin{matrix}\text{probe}\\\text{attenuation}\\\text{factor}\end{matrix}$$

and then substitute the given values

$$\text{volts peak-to-peak} = 4.6 \times 0.5 \times 10 = 23 \text{ V}$$

### NOTE

If the voltage being measured is a sine wave, the peak-to-peak value can be converted to peak, RMS, or average, as shown in Fig. 1-10. Similarly, if a peak, RMS, or average value is given and must be measured on an oscilloscope, Fig. 1-10 can be used to find the corresponding peak-to-peak value.

## 1-8. Measuring Instantaneous Voltage with an Oscilloscope

The oscilloscope is the logical tool for measurement of instantaneous or transient voltages. The following procedures describe the general steps necessary for such measurement.

1. Connect the equipment as shown in Fig. 1-11.
2. Place the oscilloscope in operation as described in the instruction manual.
3. On a laboratory oscilloscope, set the vertical step-attenuator to a deflection factor that will allow the expected signal to be displayed without overdriving the vertical amplifier.
4. On a shop oscilloscope, set the vertical gain control to the "calibrate-set" position, as determined during the calibration procedure.
5. Set the input selector (if any) to ground.

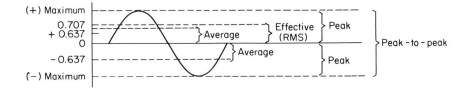

| Given ↓ | Average | Effective (RMS) | Peak | Peak - to - peak |
|---------|---------|-----------------|------|------------------|
| Average | — | 1.11 | 1.57 | 1.271 |
| Effective (RMS) | 0.900 | — | 1.411 | 2.831 |
| Peak | 0.637 | 0.707 | — | 2.00 |
| Peak - to - peak | 0.3181 | 0.3541 | 0.500 | — |

**Fig. 1-10.** Relationship of average, effective (RMS), peak, and peak-to-peak values for a-c sine waves.

## NOTE

On most laboratory oscilloscopes the input switch that selects either a-c or d-c measurement also has a position connecting the input of the vertical amplifier stage to ground. If no such switch position is provided, short the vertical input terminals by connecting the probe (or other lead) to ground.

6. Switch on the oscilloscope internal recurrent sweep. Adjust the horizontal gain control to spread the trace over as much of the screen as desired.

7. Using the vertical position control, position the trace to a line of the graticule below the center line, as shown in Fig. 1-11. This establishes the reference line. If the average signal (ac plus dc) is negative with respect to ground, position the trace to a reference line above the graticule center line. Do not move the vertical position control after this reference line has been established.

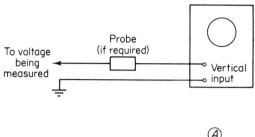

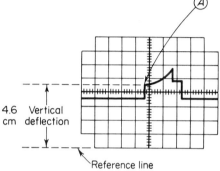

**Fig. 1-11.** Measuring instantaneous (or d-c) voltage on an oscilloscope.

## NOTE

To measure a voltage level with respect to a voltage other than ground, make the following changes. Set the input selector to measure dc. Apply the reference voltage to the vertical input; then position the trace to the reference line.

8. Set the input selector (if any) to dc. (The oscilloscope must be capable of d-c measurement. Some shop-type oscilloscopes do not have this capability.) The ground reference line, if used, can be checked at any time by switching the input selector to the ground position. Connect the probe to the signal being measured.

9. If the waveform is outside the viewing area, set the vertical step-attenuator (if any) so that the waveform is visible. Do not move the vertical position or vertical gain controls to bring the waveform into view.

10. Adjust the sweep frequency and horizontal gain controls to display the desired waveforms.

11. Measure the distance in centimeters between the reference line and the point on the waveform at which the d-c level is to be measured. For example, in Fig. 1-11 the measure is made between the reference line and point *A*.

12. Establish polarity of the signal. Any signal-inverting switches on the oscilloscope must be in the normal position. If the waveform is above the reference line the voltage is positive; if it is below the line the voltage is negative.

13. On a laboratory oscilloscope, multiply the distance measured in Step 11 by the vertical attenuator switch setting. Multiply this result by the attenuation factor of the probe, if any. This gives the instantaneous voltage.

14. On a shop oscilloscope, multiply the distance measured in Step 11 by the calibration factor (in volts per centimeter) established during calibration. Also include the attenuation factor of the probe (if any) and the setting of the attenuator step switch (if any).

For example, assume that the vertical distance measured is 4.6 cm (Fig. 1-11). The waveform is above the reference line when a 10× attenuator-probe and a vertical-deflection factor of 2 V/cm are used. Using the equation

$$\begin{matrix}\text{instantaneous} \\ \text{voltage}\end{matrix} = \begin{matrix}\text{vertical} \\ \text{distance} \\ \text{(in cm)}\end{matrix} \times \text{polarity} \times \begin{matrix}\text{volts-per-centimeter} \\ \text{factor}\end{matrix} \times \begin{matrix}\text{probe} \\ \text{attenuation} \\ \text{factor}\end{matrix}$$

we substitute the given values

$$\text{instantaneous voltage} = 4.6 \times +1 \times 2 \times 10$$

The instantaneous voltage is +92 V.

## 1-9. Voltage Measurements with a
## Variable Calibrating Source

When an oscilloscope has an internal *variable* calibrating voltage source or an accurate external source is readily available, it is often convenient to measure voltages using the variable calibration. This method is sometimes known as *indirect voltage measurement*.

The main advantage of the indirect method is that the oscilloscope vertical amplifier and graticule screen need not be precalibrated and the vertical gain control can be set to any convenient level. The basic procedure consists of measuring the test voltage and noting the number of divisions occupied by the trace. Then the test voltage is removed and replaced by the calibrating voltage. The calibrating voltage is adjusted until it occupies the same number of divisions as the test voltage. The calibrating voltage amplitude is read from the corresponding oscilloscope control or from the meter of the external calibrator. Typical external calibrator circuits are shown in Fig. 1-12. The indirect voltage measurement is quite ac-

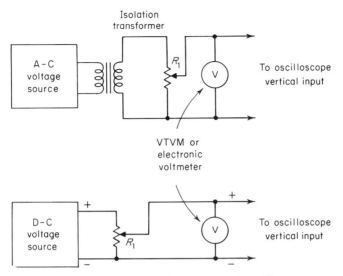

**Fig. 1-12. Typical external calibrator circuits for oscilloscopes.**

curate. However, it does require continuous switching back and forth between test and calibrating voltages.

The indirect method has some other disadvantages. If the calibrating voltage is internal, it may be fed directly to the vertical amplifier input when the appropriate control is set to a "calibrate" position. On such oscilloscopes, when a probe is used, it is necessary to include the probe multiplication factor. For example, assume that a voltage is measured through a 10-to-1 probe, that the test signal occupies three divisions, that the oscilloscope's internal square-wave calibrating signal is switched on and adjusted to occupy the same three divisions, and that the calibration signal-amplitude control indicates 7 V. Since the internal calibration voltage is applied directly to the vertical input, and the test voltage is applied through a 10-to-1 probe, the actual test voltage must be 70 V.

Another major disadvantage is that the calibrating-voltage source, internal or external, may not be equal to the test voltage. This condition can be offset by means of the vertical step-attenuator or with a voltage divider probe. For example, if the test signal reaches 300 V, whereas the calibration source has a maximum of 30 V, use a 10-to-1 voltage divider probe, or the 10× position of the vertical step-attenuator. Measure the test voltage with the probe (or the step-attenuator in 10×); then apply the calibrating voltage with the probe removed (or the step-attenuator in 1×) and adjust for the same number of divisions. Make certain to multiply the calibration voltage control indication by the appropriate factor.

## NOTE

In the following procedure, it is assumed that the oscilloscope has an internal variable square-wave calibrating voltage and that the voltage to be measured is a sine wave.

1. Connect the equipment as shown in Fig. 1-13.
2. Place the oscilloscope in operation as described in the instruction manual.
3. Set the vertical step-attenuator and/or vertical gain control to a deflection factor that will allow the expected signal to be displayed without overdriving the vertical amplifier.
4. Set the input selector (if any) to measure ac. Connect the probe to the signal being measured.
5. Switch on the oscilloscope internal recurrent sweep.
6. Adjust the sweep frequency for several cycles on the screen.
7. Adjust the horizontal gain control to spread the pattern over as much of the screen as desired.
8. Adjust the vertical position control so the downward excursion of the waveform coincides with one of the graticule lines below the graticule

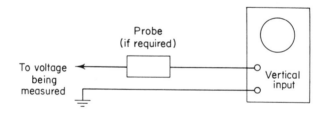

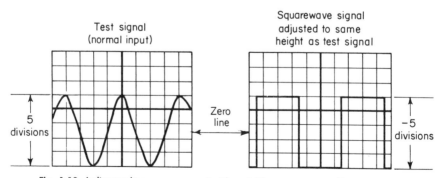

**Fig. 1-13.** Indirect voltage measurement with variable square-wave calibrating signal.

center line, as shown in Fig. 1-13. Adjust the vertical gain control so that the pattern occupies an exact, easily measured number of divisions.

9. Adjust the horizontal position control so that one of the upper peaks of the signal lies near the vertical center line, as shown in Fig. 1-13.

10. Without disturbing any of the controls, switch on the internal calibrating square-wave signals. The test signal should be removed from the screen and replaced by the square-wave calibrating signals.

11. Adjust the square-wave calibrating signal amplitude until the square-wave pattern occupies the same number of divisions as the test voltage.

## NOTE

If the square-wave signal is insufficient and cannot be adjusted to occupy the same number of divisions, switch the vertical step-attenuator to a convenient multiplier scale. Do not move the vertical gain control.

12. Read the voltage from the calibrator amplitude control. Multiply the voltage by any probe and/or step-attenuator attenuation factor.

For example, assume a peak-to-peak deflection factor, as shown in Fig. 1-13, using a $10\times$ attenuator probe. Assume, too, that the calibrator amplitude control reads 3.7 V. Further assume that the test voltage was measured with the step-attenuator in $10\times$ and the square waves were adjusted to the same height with the step-attenuator in $1\times$. We use the general equation

$$\frac{\text{volts}}{\text{peak-to-peak}} = \frac{\text{calibrator}}{\text{control}} \times \frac{\text{probe}}{\text{attenuation}} \times \frac{\text{difference in}}{\text{step-attenuator}}$$
$$\text{reading} \qquad \text{factor} \qquad \text{factor}$$

and then substitute the given values

$$\text{volts peak-to-peak} = 3.7 \times 10 \times 10 = 370 \text{ V}$$

## 1-10. Measuring Composite and Pulsating Voltage with an Oscilloscope

In practice, most voltages measured are composites of ac and dc, or are pulsating dc. For example, a transistor amplifier used to amplify an a-c signal will have both ac and dc on its collector. The output of a rotating d-c generator or solid-state rectifier will be pulsating, even though its polarity is constant. Such composite and pulsating voltages can be measured quite readily on an oscilloscope capable of measuring dc (as are most laboratory oscilloscopes).

The procedures are essentially a combination of peak-to-peak measurements (Section 1-7) and instantaneous d-c measurements (Section 1-8). Composite and pulsating voltages can also be measured by the indirect method (Section 1-9). However, this is usually quite difficult.

1. Connect the equipment as shown in Fig. 1-14.

2. Place the oscilloscope in operation as described in the instruction manual.

3. Set the vertical step-attenuator to a deflection factor that will allow the expected signal, plus any dc, to be displayed without overdriving the vertical amplifier.

4. Set the input selector to ground.

### NOTE

On most laboratory oscilloscopes, the input switch that selects either a-c or d-c measurement also has a position that connects the input

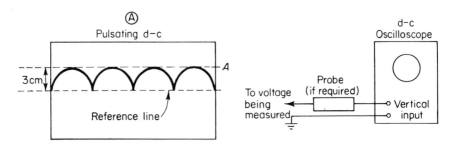

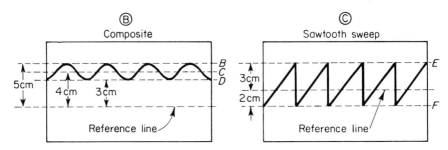

**Fig. 1-14.** Measurement of composite and pulsating voltages with an oscilloscope.

of the vertical amplifier stage to ground. If such a switch position is not provided, short the vertical input terminals by connecting the probe (or other lead) to ground.

5. Switch on the oscilloscope internal recurrent sweep. Adjust the horizontal gain control to spread the trace over as much of the screen as desired.

6. Using the vertical position control, position the trace to a convenient location on the graticule screen. If the voltage to be measured is pulsating dc, the horizontal center line should be convenient. If the voltage is a composite and the average signal (ac plus dc) is positive, position the trace below the center line. If the average is negative, position the trace above the center line. *Do not move* the vertical position control after this reference line has been established.

7. Set the input selector switch to dc. The ground reference line can be checked at any time by switching the input selector to the ground position. Connect the probe to the signal being measured.

8. If the waveform is outside the viewing area, set the vertical step-attenuator so that the waveform is visible.

9. Adjust the sweep frequency and horizontal gain controls to display the desired waveform.

10. Establish polarity of the signal. Any signal-inverting switches on the oscilloscope must be in the normal position. If the waveform is above the reference line the voltage is positive. If the waveform is below the line the voltage is negative. Measure the distance in centimeters between the reference line and the point on the waveform at which the level is to be measured.

## NOTE

If the voltage to be measured is pulsating dc, the trace will remain on one side of the reference line but will start and stop at the reference line, as shown in Fig. 1-14a. If the voltage is a composite of ac and dc, the trace may be on either side of the reference line. It is possible that the trace might cross over the reference line, but it usually remains on one side, as shown in Fig. 1-14b. If the voltage is a nonsine wave (such as sawtooth, pulse, spike, etc.), the trace may appear on both sides of the reference line, or may be displaced above or below the line, as shown in Fig. 1-14c.

11. Multiply the distance measured in Step 10 by the vertical attenuator switch setting. Also include the attenuation factor of the probe, if any.

For example, assume that the vertical distance measured is 3 cm from the reference line to point $A$ of Fig. 1-14a. The waveform is above the reference line (pulsating dc), using a $10\times$ attenuator probe and a vertical deflection factor of 2 V/cm.

Substituting the given values, we obtain

peak of the pulsating d-c voltage $= 3 \times +1 \times 10 \times 2 = +60$ V (peak)

For example, assume that the vertical distance measured is 3 cm from the reference line to point $D$ (Fig. 1-14b), 4 cm to point $C$ (Fig. 1-14b), and 5 cm to point $B$ (Fig. 1-14b). The waveform is above the reference line (ac combined with positive dc), using a $10\times$ attenuator probe and a vertical deflection factor of 2 V/cm.

Substituting the given values, we obtain

$$\text{d-c component (reference line to point } C) = 4 \times +1 \times 10 \times 2$$
$$= +80 \text{ volts}$$

$$\text{peak-to-peak of a-c component (point } B \text{ to } D) = 2 \times 2 \times 10$$
$$= 40 \text{ volts (peak-to-peak)}$$

### NOTE

The 2-cm value is obtained by subtracting the point $D$ value (3 cm) from the point $B$ value (5 cm).

For example, assume that the vertical distance measured is 3 cm from the reference line to point $E$ (Fig. 1-14c) and 2 cm from the reference line to point $F$. The waveform is above and below the reference line (sawtooth sweep), as found by using a $10\times$ attenuator probe and a vertical deflection factor of 2 V/cm. Substitute the given values

$$\text{positive peak of sweep (point } E) = 3 \times +1 \times 2 \times 10$$
$$= +60 \text{ volts (peak)}$$

$$\text{negative peak of sweep (point } F) = 2 \times -1 \times 2 \times 10$$
$$= -40 \text{ volts (peak)}$$

$$\text{peak-to-peak of sweep (point } E \text{ to } F) = 60 + 40$$
$$= 100 \text{ volts (peak-to-peak)}$$

## 1-11. Measuring Complex Voltages with a Meter

An oscilloscope is the best instrument for measuring complex voltages, since the waveshape, frequency, and peak-to-peak voltage can be found simultaneously. However, it is possible to measure the voltage of a complex wave with a meter. The best results will be obtained from a peak-to-peak reading electronic voltmeter since peak-to-peak voltage is independent of actual waveform. Other values of a complex wave (average, RMS, etc.) are dependent upon the waveform (square wave, pulse, sawtooth, half-wave pulsating, full-wave pulsating, etc.).

Most service manuals (such as those used in TV) specify the peak-to-peak value of complex waves, and so most electronic voltmeters have peak-to-peak reading scales.

Sometimes, it is convenient to convert peak-to-peak voltage of a complex wave into RMS values. *If the waveshape is known,* the peak-to-peak value can be converted into RMS using the data of Fig. 1-15.

For example, assume that a sawtooth wave is measured with an electronic voltmeter, and a peak-to-peak reading of 3 V is obtained. The RMS value of this voltage is 0.8655 (3 × 0.2885).

It is also possible to obtain a reading of a complex wave with the RMS

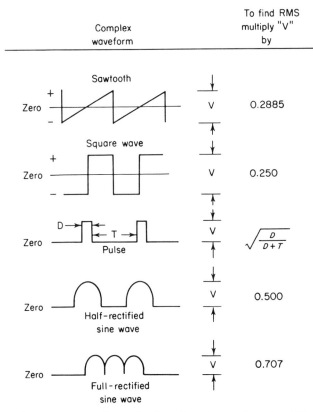

**Fig. 1-15.** Converting true peak-to-peak values of complex waveforms into RMS values.

scales of a meter. However, the indicated RMS value is not the true RMS value (because the scales are based on RMS of sine waves).

The RMS values obtained for complex waves on a conventional meter can be converted to peak-to-peak readings using the data of Fig. 1-16.

For example, assume that a square wave is measured with a VOM and an RMS reading of 3 V is obtained. The peak-to-peak value of this voltage is 5.4 (3 $\times$ 1.8).

This value is developed as follows. The average value of a full-wave rectified square wave is equal to the peak value (or one-half of the peak-to-peak value). A conventional RMS meter responds to the average value but indicates RMS, which is 1.11 times the average in sine waves. Therefore, the meter reads average (or peak in the case of a square wave) but indicates this value as 1.11 times the peak. To find the peak it is necessary to find the reciprocal of 1.11, or 0.9. To find the peak-to-peak value, multiply the peak value by two, or 2 $\times$ 0.9 = 1.8.

## 1-12. Measuring Radio Frequency Voltages

### 1-12.1. RF Probes

When the voltages to be measured are at radio frequencies and are beyond the frequency capabilities of the meter circuits or oscilloscope amplifiers, a radio frequency (or RF) probe is required. Such probes convert (rectify) the RF signals into a d-c output voltage which is almost equal to the peak RF voltage. The dc output of the probe is then applied to the meter or oscilloscope input and is displayed as a voltage readout in the

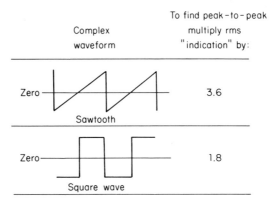

**Fig. 1-16.** Converting "indicated" values of complex waves on RMS meters into peak-to-peak values.

normal manner. In some RF probes, the meter or oscilloscope reads peak RF voltage, while in other probes the readout is RMS voltage.

No special precautions need be observed when making voltage measurements with an RF probe except to note whether the readout is in peak or RMS values (or in rare cases, peak-to-peak values).

The basic circuit of an RF probe is shown in Fig. 1-17. This circuit can be used to provide either a peak output or RMS output, but not a peak-to-peak output.

Capacitor $C_1$ is a high-capacitance, dc blocking capacitor used to protect diode $CR_1$. Usually, a germanium diode is used for $CR_1$ that rectifies the RF voltage and produces a d-c output across $R_1$. In some probes $R_1$ is omitted so that the d-c voltage is developed directly across the input circuit of the meter or oscilloscope. This d-c voltage is equal to the peak RF voltage, less whatever forward drop exists across the diode $CR_1$.

When it is desired to produce a d-c output voltage equal to the RMS of the RF voltage, a series-dropping resistor (shown in dotted form on Fig. 1-17 as $R_2$) drops the d-c output voltage to a level that equals 0.707 of the peak RF value.

### 1-12.2. Thermocouple RF Meters

RF voltages are sometimes measured by means of thermocouples. Whenever the junction formed by two dissimilar metals is exposed to a temperature difference, a voltage will be generated that is dependent on that temperature difference (and also on the temperature levels and the materials involved). This principle can be used to measure RF voltages. Usually, a *vacuum thermocouple* is used for RF voltage measurement. A vacuum thermocouple (also known as a thermal converter) is a thermojunction (two dissimilar metals) and a heater wire mounted in a small, highly evacuated glass bulb. The current (or voltage) to be measured (of

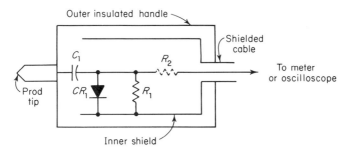

Fig. 1-17. Typical RF probe for meter or oscilloscope.

any frequency) is passed through the heater, causing a rise in temperature of the heater wire and generating a d-c voltage in the thermojunction. This voltage is proportional to the temperature rise. The output generated can be measured on a suitable millivoltmeter. Within wide limits, the accuracy of the reading obtained is independent of frequency.

Vacuum thermocouples are most often used for accurate measurements of small, high-frequency currents and voltages, calibration of highly accurate a-c standards, accurate a-c measurements at low power levels, and similar applications.

Standard vacuum thermocouples are available in either *contact* or *insulated heater* types. In the contact type (Fig. 1-18) the thermojunction is in direct electrical contact with the heater wire. The insulated heater type (Fig. 1-19) has a tiny insulating bead between the heater and thermojunction. This bead insulates the two circuits electrically but maintains good thermal contact.

Generally, the contact type is used with frequencies up to about 20 MHz, while the insulated heater types are used with frequencies from 20 MHz to 50 MHz. For frequencies above 50 MHz an ultrahigh-frequency (UHF) type of vacuum thermocouple is used. These UHF vacuum thermocouples are specifically designed for use in the accurate measurement of current and voltage at frequencies extending into UHF range. In addition to the heater and thermocouple junction being insulated from each other, the heater and thermocouple leads are brought out at opposite ends of the glass bulb (Fig. 1-20). Also, the vertical planes of the heater and junction are at right angles to each other.

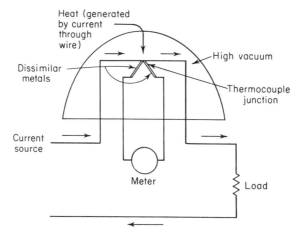

**Fig. 1-18. Contact type thermocouple meter circuit.**

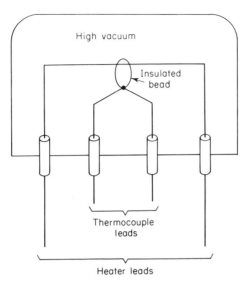

**Fig. 1-19.** Insulated heater type thermocouple junction.

Thermocouple meters are particularly useful in measuring RF voltages of complex waveforms. This is because the readout is dependent upon the heat produced by the voltage to be measured, not on the exact waveform. Thermocouple meters can be made to produce a true RMS readout of any waveform (dc, RF, complex wave, etc.) because a direct current flowing through a resistance or wire produces heat, as does an alternating current of any form (RF, complex wave, etc.). The *effective* value of an alternating current or voltage is that value which will produce the same amount of heat in a resistance as direct current or voltage of the same numerical value.

No special precautions need be observed when making voltage measurements with a thermocouple meter except to note the readout value (usually RMS).

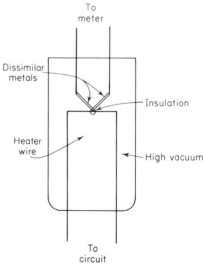

**Fig. 1-20.** UHF thermocouple construction.

## 1-13. Measuring 3-phase Circuit Voltages

The procedures for measuring 3-phase voltages (and currents) are essentially the same as for single-phase. However, it must be remembered that the voltage (or current) for each phase is not necessarily the same as for the line. Also, the output does not equal the input in a delta-to-star (or star-to-delta) transformer circuit. The important relationships in 3-phase circuits are summarized in Fig. 1-21.

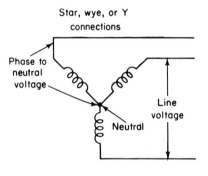

Star, wye, or Y connections

Phase to neutral voltage

Line voltage

Neutral

$$\text{Phase-to-neutral voltage} = \frac{\text{Line voltage}}{1.73}$$

Line voltage = Phase-to-neutral voltage x 1.73

Power = Phase-to-neutral voltage x 1.73 x phase current x cos $\theta$

Line current = Phase current

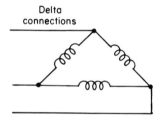

Delta connections

Line voltage = Phase voltage

$$\text{Phase current} = \frac{\text{Line current}}{1.73}$$

Line current = Phase current x 1.73

Power = Phase current x 1.73 x phase voltage x cos $\theta$

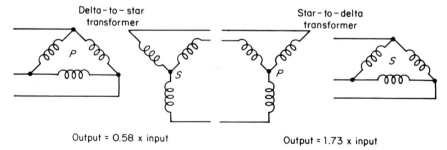

Delta-to-star transformer

Output = 0.58 x input

Star-to-delta transformer

Output = 1.73 x input

**Fig. 1-21.** Summary of three-phase voltage and current measurements.

## 1-14. Measuring Voltage-Sensitive Circuits

Many circuits are voltage sensitive. That is, the voltage will vary with changes in load. These circuits are difficult to measure since the load presented by the meter will change the voltage from the normal operating value. Since a VOM has a lower input resistance than an electronic voltmeter, the VOM presents a greater load on the circuit and thus changes the voltage by a greater amount. This is one of the advantages of an electronic voltmeter. In some cases, even an electronic voltmeter will cause circuit loading.

### 1-14.1. Determining Voltage Sensitivity

The fact that a circuit is voltage sensitive can be determined easily using the test setup of Fig. 1-22. This test setup can be applied to any circuit. The AGC line of a receiver is a typical voltage-sensitive circuit.

1. Measure the voltage directly as shown in Fig. 1-22a.

2. Insert the series resistor $R$ and measure the voltage at the same point in the circuit as shown in Fig. 1-22b. The value of resistor $R$ should be equal to the input resistance of the meter.

3. The second voltage indication (with resistor $R$ in series) should be *approximately* one-half of that obtained by direct measurement. If the second voltage indication is *much greater* than one-half, the circuit is voltage sensitive. Either of the following two methods can be used to measure voltage-sensitive circuits.

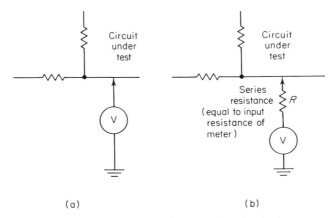

(a)                                        (b)

**Fig. 1-22.** Determining voltage sensitivity of a circuit.

### 1-14.2. Potentiometer Method

1. Connect the equipment as shown in Fig. 1-23.

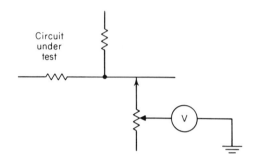

**Fig. 1-23.** Determining true voltage of a voltage-sensitive circuit (potentiometer method).

2. Set the potentiometer to zero ohms and measure the circuit voltage. This will be the initial voltage.

3. Increase the potentiometer resistance until the voltage is one-half of the initial voltage (one-half of that obtained in Step 2).

4. Disconnect the potentiometer from the circuit. Measure the d-c resistance of the potentiometer.

5. Find the true circuit voltage with the following equation

$$\text{true voltage} = \frac{\text{initial voltage} \times \text{potentiometer resistance}}{\text{meter input resistance}}$$

For example, assume that the initial circuit reading was 3 V, the potentiometer resistance needed to reduce the reading to 1.5 V was 140,000 ohms, and the meter input resistance was 60,000.

$$3 \times 140,000 = 420,000; \qquad \frac{420,000}{60,000} = 7\text{ V}$$

Therefore, the true circuit voltage is 7 V, and the meter caused a 4-V drop.

### 1-14.3. Opposing Voltage Method

1. Connect the equipment as shown in Fig. 1-24a. The value of series resistance $R$ is not critical but should be near that of the meter input resistance.

2. Adjust the opposing voltage source until the meter reads zero.

3. Disconnect the meter and opposing voltage source.

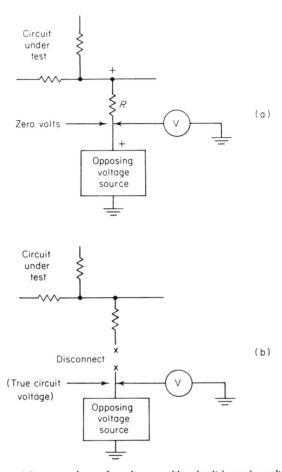

**Fig. 1-24.** Determining true voltage of a voltage-sensitive circuit (opposing voltage method).

4. Reconnect the meter to measure the opposing voltage source as shown in Fig. 1-24b.

5. The true circuit voltage is equal to the opposing source voltage.

# CHAPTER 2

# Current Measurements

## 2-1. Kirchhoff's Current Law

In some applications, it is often convenient to solve a network by means of *Kirchhoff's current law*.

In any electrical network, the algebraic sum of the currents that meet (entering or leaving) at a point is zero; or, the algebraic sum of the currents flowing toward a junction is equal to zero.

Fig. 2-1 shows the relationship of the currents in a typical circuit.

## 2-2. Ohm's Law for Direct Current

Ohm's law formulas are used more widely than any other in electronics. The basic Ohm's law formulas for direct current are given by

$$E = IR, \qquad I = E/R, \qquad R = E/I, \qquad P = EI$$

where $I$ is current in amperes,
  $R$ is resistance in ohms,
  $E$ is potential across $R$ in volts, and
  $P$ is power in watts.
The calculations for Ohm's law (dc) are shown in Fig. 2-2.

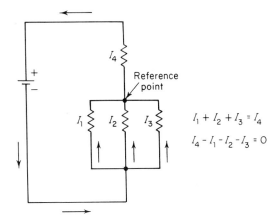

Fig. 2-1. Example of Kirchhoff's currrent law.

$$I_1 + I_2 + I_3 = I_4$$

$$I_4 - I_1 - I_2 - I_3 = 0$$

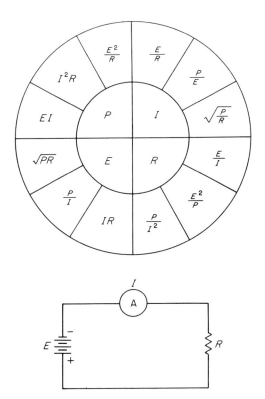

Fig. 2-2. Calculations for d-c Ohm's law.

## 2-3. Ohm's Law for Alternating-Current Circuits

Ohm's law formulas for a-c circuits are essentially the same as for d-c circuits except that impedance $(Z)$ is substituted for resistance $(R)$. Therefore,

$$E = IZ, \qquad I = E/Z, \qquad Z = E/I, \qquad P = IE$$

where  $I$ is current in amperes,

$Z$ is impedance in ohms,

$E$ is potential across $Z$ in volts, and

$P$ is *apparent power* supplied by the source in voltamperes.

The calculations for Ohm's law (ac) are shown in Fig. 2-3.

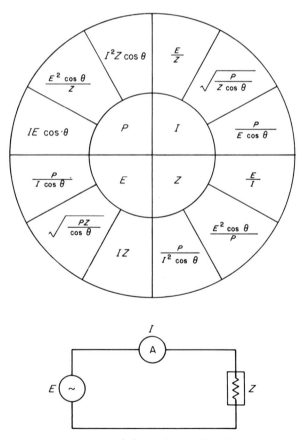

**Fig. 2-3.** Calculations for a-c Ohm's law.

The term *apparent power* is applied when the reactive power factor of an a-c circuit is considered. To find the *true* power in a series a-c circuit it is necessary to multiply the apparent power by the cosine of the phase angle between voltage and current. (Methods for determining phase angle in a-c circuits are discussed in Chapter 8.) Since the phase angle is always less than 90° and the corresponding cosine is a fraction of one, the true power will always be less than apparent power.

The ratio between the true power and the apparent power is known as the *power factor*. This can be expressed as

$$\text{power factor} = \frac{E \times I \times \text{cosine phase angle}}{E \times I}$$

or

$$\text{power factor} = \text{cosine phase angle} = \cos \frac{R}{Z}$$

where $R$ is resistance in an a-c circuit (in ohms) and
$Z$ is impedance (in ohms).

In those a-c circuits where there is no reactance (pure resistance) the phase angle is zero; therefore, apparent power is the true power. (Reactance and impedance are discussed further in Chapters 5 and 6.)

## 2-4. Basic Ammeter (Current) Measurements

The following paragraphs describe the *basic* steps necessary to measure current with a volt-ohmmeter (VOM).

Most electronic voltmeters do not have a provision for measuring current primarily because of their high input impedance. Since current must pass through the meter input circuit, there is a voltage drop across the meter. In an electronic meter the voltage drop could be very high.

The first step in making a current measurement is to set the range. Always use a range that is *higher* than the anticipated current. If the approximate current is not known, use the highest current range initially and then select a lower range, insuring a good mid-scale reading.

In many meters, selecting a current range involves more than positioning a switch. A typical VOM may require that the test leads be connected to different terminals from those used for voltage measurement.

Next, set the function selector to ac or dc as required. Most VOMs will not measure a-c, so either plus or minus dc must be selected.

Note that when the lowest current scale is selected, such as 50 $\mu$A, the meter is actually functioning as a voltmeter. The meter movement is placed in series with the circuit without a shunt. Therefore, any sudden surge of current can damage the meter movement. This is a problem especially where there is both ac and dc in the circuit being measured. If the ac is of a higher frequency, it will probably have little effect on the meter movement. Lower frequency ac can combine with the dc and possibly cause reading errors or meter movement burnout.

Remember that all current measurements (ac, dc, plus, minus) are made with the meter in *series* with the circuit and power source, as shown in Fig. 2-4. This means that all of the current normally passing through the circuit under test will be passed through the meter. This may or may not affect circuit operation.

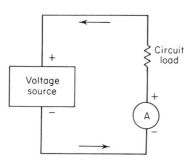

**Fig. 2-4. Current measurements are made with meter in series with circuit and voltage source.**

## 2-5. Checking Ammeter Accuracy

The scales of an ammeter must be checked for accuracy of the current indication. A typical VOM will have only d-c ranges, while a typical electronic voltmeter will have no current ranges. Therefore, the primary concern is with d-c accuracy. However, if a meter is provided with a-c scales, they must also be checked for accuracy over the entire rated frequency range. Since this handbook is concerned primarily with test and measurement procedures, and assumes that the reader is familiar with test equipment calibration, no detailed current calibration procedures are given here. The author's *Handbook of Electronic Meters: Theory and Application* (Englewood Cliffs, N.J.: Prentice-Hall, Inc., 1969) provides a comprehensive discussion of methods for current calibration. The following paragraphs describe *basic* procedures for checking ammeter accuracy.

### 2-5.1. Current Accuracy

The most convenient method of checking current accuracy is to compare the ammeter to be tested against a standard ammeter of known accuracy. The ammeter to be tested is connected in series with the standard ammeter and a variable current source, as shown in Fig. 2-5. The source is then varied over the entire range of the ammeter, and the current indications are compared. The accuracy of this test is dependent entirely on the accuracy of the standard meter, not on the current source.

The next most popular method of checking current accuracy is with a precision voltmeter and precision resistance. The precision resistance and ammeter to be checked are connected in series with a variable power source as shown in Fig. 2-6. The precision voltmeter is connected across the precision resistance. Current through the circuit is computed by Ohm's law $(I = E/R)$.

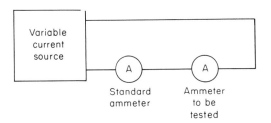

Fig. 2-5. Basic ammeter calibration circuit.

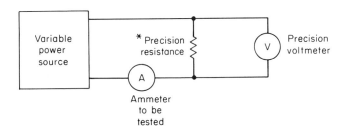

\* For a 1-ohm value, 1 V = 1 A

For a 1000-ohm value, 1 V = 1 mA

Fig. 2-6. Ammeter calibration circuit using a precision resistance and precision voltmeter.

The value of the precision resistor is chosen so that the voltage indicated on the precision voltmeter can be related directly to current. For example, if a 1-ohm resistance is used, the voltage across the resistance can be read directly in amperes (3 V equals 3 A, 7 V equals 7 A, etc.). If a 1000-ohm precision resistor is used, the voltage can be read directly as milliamperes (3 V equals 3 mA, etc.).

Note that the accuracy of this test method is dependent upon the accuracy of both the voltmeter and series resistance. The tolerance of both components must be added. For example, assume that both the voltmeter and resistance have a 1% tolerance. Then accuracy of the circuit could be no greater than 2%. In any event, the combined accuracy must be greater than that of the meter to be tested.

## 2-6. Extending Ammeter Ranges

The range of an ammeter can be extended to measure high current values by using *external* shunts connected to the basic meter movement (usually the lowest current range), as shown in Fig. 2-7. Values for shunts can be determined using the following procedures.

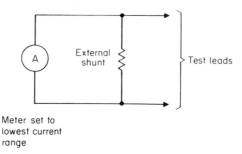

Meter set to
lowest current
range

**Fig. 2-7.** Extending ammeter ranges with an external shunt.

## NOTE

The range of the basic meter movement cannot be lowered. For example, if a 100-$\mu$A movement with 100 scale divisions is used to measure 1 $\mu$A, the meter will deflect only one division. It is not practical to obtain greater deflection.

### 2-6.1. Determining Shunt Resistance Values

The shunt resistance required to convert a basic meter movement into an ammeter capable of measuring a given current range can be calculated in either of two ways.

*Method No. 1.* Use the following equation

$$R_x = \frac{R_m}{N - 1}$$

where $R_x$ is the shunt resistance (in parallel with the meter movement),
   $R_m$ is the internal resistance of the meter movement (refer to Section 1-4.1 for procedure to find internal resistance of meters), and
   $N$ is the multiplication factor by which the scale factor is to be increased.

For example, assume that a meter has a 50-$\mu$A full-scale movement (with 50 equal divisions on the scale), an internal resistance of 300 ohms, and it is desired to convert the meter to measure 0 to 100 mA. (Each of the 50 scale divisions will then represent 2 mA.)

1. Find the multiplication factor (or $N$ factor) by which the movement is to be increased; $N$ = desired scale factor (0.1 A, or 100 mA) divided by movement current (0.00005 A), or 2000.

2. Using the equation

$$R_x = \frac{300}{2000 - 1} = 0.1508 \text{ ohm}$$

*Method No. 2.*

1. Find the voltage required for full-scale deflection of the meter movement

$$E = IR \quad \text{or} \quad (5 \times 10^{-5})(3 \times 10^2) = 0.015 \text{ V}$$

2. Subtract the meter movement full-scale current (0.00005) from the desired current (0.1) to find the current that must flow through the shunt

$$0.1 - 0.00005 = 0.09995 \text{ A}$$

3. Using Ohm's law, find the shunt resistance

$$R = \frac{E}{I} \quad \text{or} \quad \frac{0.015}{0.09995} = 0.1508 \text{ ohm}$$

### 2-6.2. Fabricating Temporary Shunts

The internal shunts of commercial meters are precision resistors. External shunts are usually strips or bars of metal connected directly between the meter movement terminals. Sometimes shunts are strips of metal mounted on insulators. Commercial shunts should be used for permanent operation. However, for *temporary use,* it is possible to extend the range of any basic meter movement many times by using nothing more than a piece of wire.

The basic arrangement is shown in Fig. 2-8, while the calibration schematic is shown in Fig. 2-9. If a piece of wire is added across the terminals of a basic movement, part of the current will pass through the wire. If the wire is twisted as shown it is possible to adjust the wire's resistance, thereby controlling the amount of current flowing through the wire.

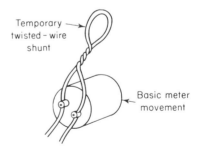

Fig. 2-8. Basic arrangement for temporary twisted-wire shunt.

The wire is twisted or untwisted as necessary, depending on whether more or less resistance is need. It is not necessary to know the resistance of the shunt or the internal resistance of the meter movement, only the full-scale deflection of the movement and the full-scale deflection that is desired.

1. Connect the meter movement to the potentiometer and voltage source as shown.

2. Set the potentiometer to its full-scale value *before* connecting the meter. Then gradually reduce the potentiometer resistance until the meter reads full scale, or 50 $\mu$A.

3. Connect the twisted-wire shunt. The meter movement should drop back towards zero.

4. Twist or untwist the wire until the meter movement reads exactly half scale, or 25 $\mu$A. (The actual current will still be 50 $\mu$A, since the potentiometer resistance is not changed.)

5. With the shunt wire connected and twisted so that an actual 50-$\mu$A flow will show as 25 $\mu$A (half scale), then full scale will be 100 $\mu$A. Thus, the 50-$\mu$A movement has been extended to 100 $\mu$A.

6. If it is desired to extend the range further, adjust the potentiometer until the meter again reads full scale (which is now 100 $\mu$A). Then twist or untwist the shunt wire until the movement again reads half scale. The actual current flow will still be 100 $\mu$A, and the meter will be extended to a 200-$\mu$A full-scale rating.

7. In theory, the process could be repeated any number of times until the meter movements were extended to any desired range. In practice, a twisted-wire shunt has some obvious drawbacks. If the wire is exposed to any handling the shunt resistance will change and throw the calibration off, but the method is quite accurate for temporary use.

## 2-7. In-Circuit Current Measurements

In-circuit current measurements are always a problem since the circuit must be interrupted (unless a clamp-on adapter or current probe is used). Alternating-current measurements are a particular problem since most VOMs do not measure ac beyond 60 Hz, while electronic meters do not measure any form of current. Two basic solutions to the problem are measurement with a test resistance and measurement with a transformer.

### 2-7.1. Current Measurement with a Test Resistance

Both ac and dc can be measured by inserting a low-value noninductive resistance in series with the circuit, measuring the voltage across the resistance, and then calculating the current using basic Ohm's law ($I = E/R$), as shown in Fig. 2-10. If a 1-ohm resistor is used, the voltage indication can be converted directly into current (3 V = 3 A, 7 mV = 7 mA, etc.).

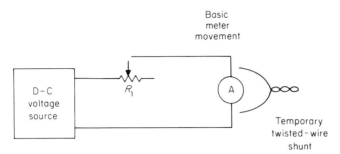

**Fig. 2-9.** Calibration circuit for twisted-wire shunt.

### 2-7.2. Current Measurement with a Transformer

Alternating current can be measured by means of a transformer as shown in Fig. 2-11. The current to be measured is passed through the transformer primary. The voltage developed across the secondary and lead is measured by the meter (or oscilloscope).

There are commercial versions of this arrangement used in heavy power equipment. These are known as "current transformers" or "current adapters." In the commercial equipment, there are several taps on the primary winding. Each tap is designed to carry a particular current and produce a given voltage across the load. The commercial current transformer circuit can be duplicated with individual components. However, because of the calibration problem (producing a specific voltage for a given current) it is more practical to use commercial equipment. Therefore, the method of measuring alternating current shown in Fig. 2-10 is

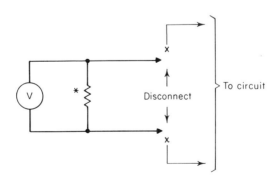

*1-ohm noninductive resistor, 1 V = 1 A

**Fig. 2-10.** Measuring in-circuit current with resistor and voltmeter.

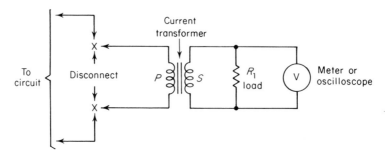

**Fig. 2-11.** Measuring in-circuit alternating current with transformer and voltmeter.

recommended where commercial current adapters or transformers are not available.

## 2-8. Current Measurement with a Current Probe

Measuring current with a test resistor or transformer as described in Section 2-7 has two disadvantages. First, the circuit must be opened so that the resistor or transformer can be inserted during the test. Second, operation of the circuit can be affected by the additional resistance. Both of these problems can be eliminated by means of a *current probe*. Such probes can be obtained as accessories for most laboratory oscilloscopes and meters.

Current probes operate on the same basic principle as the clamp-type ammeters used in power electrical equipment. The basic element of a current probe is a ferrite core with a handle. Ferrite material is used to provide a wide frequency response. The core, shown functionally in Fig. 2-12, is designed to be opened and closed so that it can be clamped around the wire carrying the current to be measured. The wire forms the

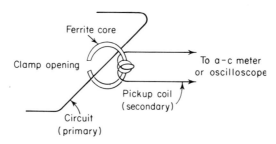

**Fig. 2-12.** Basic clip-on meter or current probe circuit.

primary of a transformer. The secondary is formed by the probe pickup coil. Current passing through the wire induces a voltage in the secondary. The secondary output voltage is usually quite high in respect to the primary since the pickup coil has many turns.

The probe output voltage is applied to the oscilloscope or meter and is measured in the normal manner. Since there is a direct relationship between voltage and current, the current can be calculated from the voltage indicated on the meter or oscilloscope.

If the current probe is used with an oscilloscope and the oscilloscope vertical amplifier is calibrated for a given value of reference deflection, the current may be read directly from the oscilloscope screen. For example, if the probe output is 1 mV/mA of current and the oscilloscope vertical amplifier is calibrated from 1 mV/cm, the current may be read directly from the scale (1 mA/cm).

If the current probe is used with a voltmeter, the current may be converted directly from the voltage indication, usually on a one-to-one basis (1 V equals 1 A, 1 mV equals 1 mA, etc.).

For oscilloscope operation, current probes are often used with an amplifier since the probe output is quite low in relation to the average oscilloscope deflection factor. A typical probe has an output of 1 mV/mA, whereas the average laboratory oscilloscope may have a vertical sensitivity of 5 mV/cm. Thus, it would require 50 mA to produce a deflection of 1 cm.

Since current probes are used with laboratory oscilloscopes or meters and are provided with detailed instructions for their use, no operating procedures are given here.

## 2-9. Measuring 3-phase Circuit Currents

The procedures for measuring 3-phase currents are essentially the same as for measuring single-phase circuits. However, it must be remembered that the current for each phase is not necessarily the same as for the line. Also, the output does not equal the input in a delta-to-star (or star-to-delta) transformer circuit. The important relationships (both voltage and current) in 3-phase circuits are summarized in Fig. 1-21.

# CHAPTER 3

# Resistance Measurements

## 3-1. Basic Resistance Equations

Although resistors can be used in a variety of circuit combinations, these circuits are a version of the basic series, parallel, or series-parallel arrangements shown in Figs. 3-1 through 3-4.

## 3-2. Resistor Color Codes

Modern resistors of the molded composition type have colored encircling bands grouped at one end, as shown in Fig. 3-5. The color coding for the resistors is shown from left to right. Usually, four bands of color are present for the carbon composition types, and five bands are used for the film types. In either case, the last color denotes the tolerance that must be applied to the value obtained. Thus if a rated 100-ohm resistor has a tolerance of 10%, it could have an actual value of between 90 and 110 ohms. (The abbreviation GMV stands for *guaranteed minimum value.*) When a value is marked *alternate,* it indicates a coding that may have been used in the past but is now generally superseded in modern components by the preferred coding designations. It should be noted that not all precision resistors are color coded.

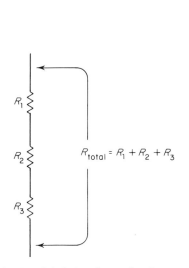

$$R_{total} = R_1 + R_2 + R_3$$

**Fig. 3-1.** Calculations for total resistance of resistors in series.

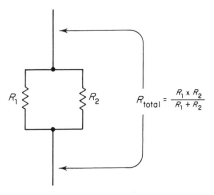

$$R_{total} = \frac{R_1 \times R_2}{R_1 + R_2}$$

$R_{total}$ is always less than the value of either resistor. If $R_1 = R_2$, then $R_{total}$ is one-half of $R_1$ or $R_2$. Where there is a 10 to 1 (or better) ratio between $R_1$ and $R_2$, $R_{total}$ will be slightly less than smallest resistor value.

**Fig. 3-2.** Calculations for total resistance of two resistors in parallel only.

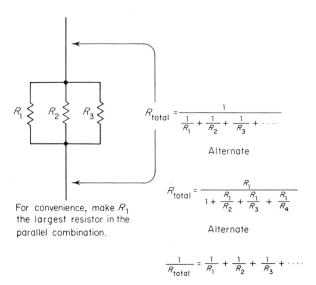

$$R_{total} = \frac{1}{\dfrac{1}{R_1} + \dfrac{1}{R_2} + \dfrac{1}{R_3} + \cdots}$$

Alternate

For convenience, make $R_1$ the largest resistor in the parallel combination.

$$R_{total} = \frac{R_1}{1 + \dfrac{R_1}{R_2} + \dfrac{R_1}{R_3} + \dfrac{R_1}{R_4}}$$

Alternate

$$\frac{1}{R_{total}} = \frac{1}{R_1} + \frac{1}{R_2} + \frac{1}{R_3} + \cdots$$

**Fig. 3-3.** Calculations for total resistance of three (or more) resistors in parallel.

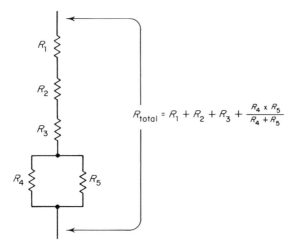

**Fig. 3-4.** Calculations for total resistance of series-parallel resistor combinations.

## 3-3. Solving Resistance Network Problems

The equations shown in Figs. 3-1 through 3-4 are sufficient to solve most d-c resistance problems. However, there are two other methods that may prove convenient in some cases.

### 3-3.1. Assumed Voltage Method

Figure 3-6 shows that a 500-ohm resistor must be placed in parallel with an unknown resistor to produce an equivalent 100 ohms.

Assume a voltage of 500 V across both resistors, producing 1 A in the known 500-ohm resistor, where $R = E/I$. The same 500 V would produce 5 A through the equivalent 100-ohm resistance. Since 5 A flow through the total network and 1 A flows through the known 500-ohm resistor, 4 A must be flowing through the unknown resistor. Since $R = E/I$, the unknown resistance must be 125 ohms ($125 = 500/4$).

### 3-3.2. Current Method

In practical application it is often convenient to find resistance values by means of currents passing through a circuit. This is particularly true for parallel-resistance networks, since the same voltage appears across each of the resistors.

Figure 3-7 shows the equations necessary to find either current or resistance in a parallel-resistance network. This applies where there are

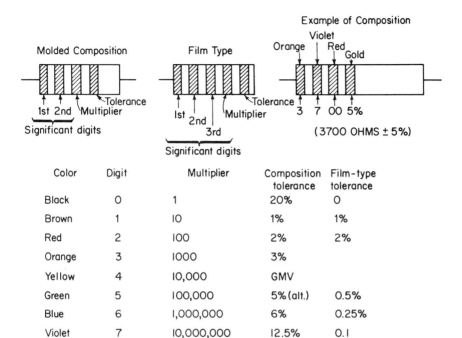

| Color | Digit | Multiplier | Composition tolerance | Film-type tolerance |
|---|---|---|---|---|
| Black | O | 1 | 20% | O |
| Brown | 1 | IO | 1% | 1% |
| Red | 2 | IOO | 2% | 2% |
| Orange | 3 | IOOO | 3% | |
| Yellow | 4 | 10,000 | GMV | |
| Green | 5 | 100,000 | 5% (alt.) | 0.5% |
| Blue | 6 | 1,000,000 | 6% | 0.25% |
| Violet | 7 | 10,000,000 | 12.5% | 0.1 |
| Gray | 8 | 0.01 (alt.) | 30% | 0.05 |
| White | 9 | 0.1 (alt.) | 10% (alt.) | |
| Silver | | 0.01 (preferred) | 10% (pref.) | 10% |
| Gold | | 0.1 (preferred) | 5% (pref.) | 5% |
| No Color | | | 20% | |

Fig. 3-5. Resistor color coding.

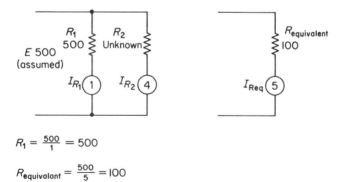

$$R_1 = \frac{500}{1} = 500$$

$$R_{equivalent} = \frac{500}{5} = 100$$

$$R_{2(unknown)} = \frac{500}{4} = 125$$

Fig. 3-6. Solving resistance network problems with assumed voltages.

two resistors and three of the four values are known. The four equations of Fig. 3-7 are based on the *shunt law* (the ratio of currents is inversely proportional to the ratio of resistances).

$$I_1 = \frac{I_2 \times R_2}{R_1} \qquad I_2 = \frac{I_1 \times R_1}{R_2}$$

$$R_1 = \frac{I_2 \times R_2}{I_1} \qquad R_2 = \frac{I_1 \times R_1}{I_2}$$

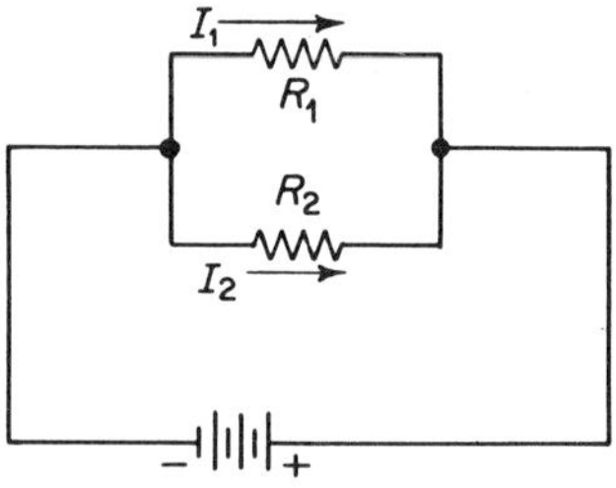

**Fig. 3-7.** Example of shunt law application to find unknown resistance and/or current values in a parallel-resistance network.

## 3-4. Basic Ohmmeter (Resistance) Measurements

The following paragraphs describe the basic steps necessary to measure resistance with an ohmmeter.

1. Zero the meter on the resistance range to be used. The meter can be zeroed on other ranges and, on some meters, will remain constant for all ranges. On other meters, the ohmmeter zero will change for each range. The meter is usually zeroed by touching the two test prods together and adjusting the ZERO OHMS, or OHMS, control until the pointer is at "ohmmeter zero." This is usually at the right end of the scale for a VOM, and at the left end for an electronic meter.

2. Once the ohmmeter is zeroed, connect the test prods across the resistance to be measured.

3. Read the resistance from the ohmmeter scale. Make certain to apply any multiplication indicated by the range selector switch. For example, if an indication of three is obtained with the range selector at $R \times 10$, the resistance is 30 ohms. It should be possible to set the range selector at $R \times 1$ and obtain a direct reading of 30 ohms. However, it may or may not be necessary to zero the ohmmeter when changing ranges.

Two major problems must be considered in making any ohmmeter measurements. First, there must be no power applied to the circuit being measured. Any power in the circuit might damage the meter and will cause an incorrect reading. Remember, capacitors often retain their charge after power is turned off. With power off, short across the circuit to be measured with a screwdriver discharging any capacitance. Then make the resistance measurement.

Second, make certain that the circuit or component to be measured is not in parallel with (shunted by) another circuit or component that will pass direct current.

The simplest method to eliminate the parallel resistance is to disconnect one lead of the resistance being measured.

## 3-5. Checking Ohmmeter Accuracy

Ohmmeter accuracy can be checked by measuring the value of precision resistors. If the indicated resistance values are within tolerance, the ohmmeter can be considered as operating properly and ready for use. The following points should be considered when making ohmmeter accuracy checks.

The resistors should have a 1% or better rated accuracy tolerance. In any event, the resistor accuracy must be greater than the rated ohmmeter accuracy. A typical ohmmeter will have a ±2% or ±3% accuracy.

Select resistor values that will provide mid-scale indications *on each ohmmeter range.* Make certain to zero the ohmmeter when changing ranges.

To determine *distributed error,* select precision test resistors that will give 25% and 75% scale indications in addition to the 50% (mid-scale) indication. Accuracy will not be the same on all parts of the scale, due to nonuniform meter movements. However, accuracy should be within the rated tolerance on all parts of the scale and on all ranges.

Often, ohmmeter accuracy will be rated in degrees of arc, rather than a percentage of full scale. It is sometimes difficult to relate the degrees of arc that a pointer will travel to a percentage. For practical work, remember that the ohmmeter accuracy is *approximately* equal to the accuracy of the meter's d-c scale. For example, assume that the d-c scale is rated as accurate to within ±2 small divisions (the scale has 100 small divisions with a ±2% accuracy). Then the ohmmeter scale will also be accurate within the same degree of pointer travel (or arc) as it takes for the pointer to move ±2 small divisions on the d-c scale.

## 3-6. Extending Ohmmeter Ranges

It is often convenient to extend the range of an ohmmeter to measure either very high or very low resistances. This is possible using high-ohms or low-ohms adapters. The circuit for a typical high-ohms adapter is shown in Fig. 3-8. This basic circuit is suitable for use with a VOM or electronic meter. A low-ohms adapter suitable for a VOM is shown in Fig. 3-9. These adapters may be packaged as a form of probe if desired. For greatest convenience, the adapter circuit values should be chosen for a multiplication factor of ten. That is, a high-ohms adapter should increase the ohmmeter reading by ten, and a low-ohms adapter should decrease the reading by ten.

Note that while both circuits use the ohmmeter scales for readout, the

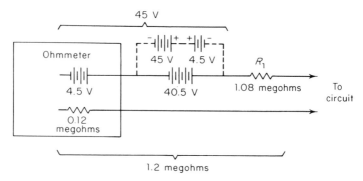

**Fig. 3-8.** High-ohms adapter circuit.

low-ohms adapter circuit does not use the VOM internal ohmmeter circuit. Instead, the adapter circuit is connected to the lowest current scale (usually this is connected directly to the meter movement).

### 3-6.1. High-ohms Adapter

To increase the ohmmeter range by a factor of ten, operate the meter at its highest ohmmeter range (usually $R \times 10,000$ or $R \times 100,000$) and connect the circuit shown in Fig. 3-8.

The external battery voltage should be such that the total voltage (external battery plus internal ohmmeter battery) is ten times that of the internal ohmmeter battery. For example, if a 4.5-V internal battery is used, the total voltage should be 45 V, and the external battery should be 40.5 V. If it is not practical to obtain a 40.5-V battery or source, a 45-V battery should be used, with a 4.5-V opposing battery in series. This arrangement is shown in Fig. 3-8. The total voltage will then be the desired 45 V.

The value of the external resistor should be such that the total re-

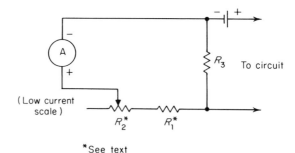

**Fig. 3-9.** Low-ohms adapter circuit.

sistance (external resistor plus input resistance on the highest ohmmeter range) is ten times that of the input resistance. A good approximation of the input resistance can be obtained by noting the center-scale indication on the ohmmeter. Typical center-scale indications on VOMs are 12 and 4.5. These represent 120,000- and 45,000-ohms input resistance respectively on the $R \times 10,000$ scale. Assuming that the input resistance is 120,000, the total resistance would then be 1.2 megohms. This would require an external resistor of 1.08 megohms (1.1 megohms for practical purposes). The exact value of the external resistor is not especially critical, since the entire circuit is adjusted for zero with the ohmmeter's ZERO OHMS control.

The high-ohms adapter is used in the same manner as any conventional ohmmeter circuit. The test leads are shorted together, and the meter is zeroed with the ZERO OHMS control. Then measurements are made in the normal manner. To check the accuracy of a high-ohms adapter, measure the value of a precision resistor. The accuracy of the high-ohms circuit readings should be within 1% of the readings made with the ohmmeter. Therefore, if the ohmmeter is rated at $\pm 3\%$ then readings obtained with the high-ohms adapter connected should be $\pm 4\%$.

### 3-6.2. Low-ohms Adapter

To decrease the ohmmeter range by a factor of ten, operate the meter at its lowest current range (usually this is at the basic meter movement and will be 50 $\mu$A for a 10,000-ohms/V meter, 100 $\mu$A for a 20,000-ohms/V meter, etc.) and connect the circuit of Fig. 3-9.

The external battery voltage is not critical, and 1.5 V is chosen for convenience. However, the battery should have a high current capacity since there will be heavy current drain in the low-ohms circuit.

The value of fixed resistor $R_1$ and external zero adjust resistor $R_2$ combined in series should be approximately ten times the internal resistance of the meter movement. The exact values of $R_1$ and $R_2$ are not critical since the entire circuit is adjusted for zero with $R_2$

The value of shunt resistor $R_3$ should be *approximately* 0.1 times the center-scale ohms value when the ohmmeter is set to its lowest resistance range (usually $R \times 1$). For example, if the center-scale indication is 12 ohms (on $R \times 1$), the value of $R_3$ should be 1.2 ohms. A more exact value would be 0.095 times the center-scale reading, or 1.14 ohms for a 12-ohm indication.

The value of $R_3$ is critical, since the shunt resistor determines the 10-to-1 scale reduction. To check the accuracy of the low-ohms adapter, short the test leads together, zero the circuit with $R_2$, and then measure the value of a precision resistor. Use a precision standard resistor in the 3-to-7 ohm range. Remember that a 3-ohm resistor will show a 30-ohm reading on the ohmmeter scale. Then try a 1-ohm (or less) precision resistor. If necessary, select a different value of $R_3$ to obtain a correct low-resistance reading.

Make certain to zero the circuit with $R_2$ before making each resistance measurement. Also, do not keep the low-ohms adapter circuit connected to the circuit under test (or hold the test leads shorted) for any length of time, as this will cause considerable current to flow through $R_3$. The heat thus generated may change the resistance value of $R_3$. Use a wire that is not heat sensitive (such as Manganin wire) for best results.

The low-ohms adapter is used in the same manner as any conventional ohmmeter circuit. The test leads are shorted together, and the meter is zeroed with the variable series resistance $R_2$. Then measurements are made in the normal manner. The low-ohms adapter is not effective in measuring values such as the resistance of a cold-soldered joint, the resistance of a switch contact, etc.

## 3-7. Resistance Bridges

Precision resistance values are often measured by means of bridge circuits. Although there are many resistance bridge circuits in use, the Wheatstone and Kelvin are best known.

### 3-7.1. Wheatstone Bridge

The basic Wheatstone bridge circuit is shown in Fig. 3-10. Resistors $R_A$ and $R_B$ are fixed and of known value. $R_S$ is a variable resistor with the necessary calibration arrangement to read the resistance value for any setting (usually a calibrated dial coupled to the variable resistance shaft). The unknown resistance value $R_X$ is connected across terminals $B$ and $C$. A battery or other power source is connected across points $A$ and $C$.

When switch $S_1$ is closed, current flows in the direction of the arrows, and there is a voltage drop across all four resistors. The drop across $R_A$ is equal to the drop across $R_B$ (provided that $R_A$ and $R_B$ are of equal resistance value). Variable resistance $R_S$ is adjusted so that the galvanometer reads zero (center scale) when switch $S_2$ is closed. At this adjustment, $R_S$ is equal to $R_X$ in resistance. By reading the resistance of $R_S$ (from the calibrated dial), the resistance of $R_X$ is known.

When the variable resistance $R_S$ is equal to $R_X$, the difference of potential between points $B$ and $D$ will be zero, and no current will flow

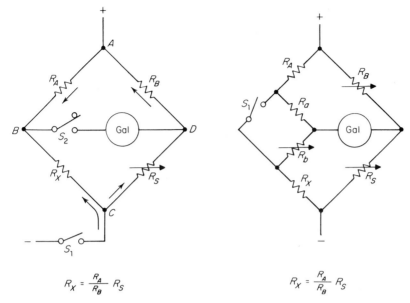

$$R_X = \frac{R_A}{R_B} R_S$$

**Fig. 3-10.** Basic Wheatstone resistance bridge.

$$R_X = \frac{R_A}{R_B} R_S$$

**Fig. 3-11.** Basis Kelvin resistance bridge.

through the galvanometer. If $R_S$ is not equal to $R_X$, then $B$ and $D$ are not at the same voltage, and current will flow through the galvanometer, moving the pointer away from zero (center scale).

The equation shown in Fig. 3-10 is used when the values of $R_A$ and $R_B$ *are not equal.* In some commercial bridges, the value of $R_A$ is ten times that of $R_B$. Thus the actual value of $R_X$ is ten times the indicated value of $R_S$. This permits a large value of $R_X$ to be measured with a low-value $R_S$. In other bridges, the value of $R_B$ is ten times that of $R_A$. Thus the actual value of $R_X$ is $\frac{1}{10}$ the indicated value of $R_S$. This permits a small value of $R_X$ to be measured with a high-value $R_S$. Commercial Wheatstone bridges can be used to make measurements from 1 ohm to 1 megohm with accuracy of $\pm 0.25\%$.

### 3-7.2. Kelvin Bridge

The basic Kelvin bridge circuit is shown in Fig. 3-11. The Kelvin bridge is similar to the Wheatstone and is used for measuring small resistances as low as 0.001 ohm with an accuracy of $\pm 2\%$, or better. The relationship of $R_a/R_b = R_B/R_S$ must be maintained for the given balance equation. This is accomplished by varying $R_S$ and $R_B$ until a null is obtained with the switch closed. The switch is then opened and $R_a$ and $R_b$ are adjusted for null. The process is repeated until a final null is obtained with the switch open or closed.

## 3-8. Testing Potentiometers and Variable Resistors

The resistance value of a potentiometer or variable resistor can be measured with an oscilloscope. In addition to checking the resistance value, it is possible to check the contact of a potentiometer for "noise" or "scratchiness" using an ohmmeter or an oscilloscope.

### 3-8.1. Ohmmeter Method

Connect the ohmmeter between the wiper contact and one end of the winding. Then rotate the contact through the full range of resistance. The ohmmeter resistance reading should vary smoothly throughout the range. The linearity of the ohmmeter reading will depend upon the linearity of the potentiometer. However, if the meter pointer "jumps" as the resistance is varied, the potentiometer contact is not riding firmly on the resistance winding or composition element. Often, this condition is the result of dirt on the contact.

### 3-8.2. Oscilloscope Method

An oscilloscope can be used to check the noise (both static and dynamic) of a potentiometer or variable resistance. Static potentiometer "noise" is a result of any current variation due to poor contact when the contact arm is at rest. Dynamic noise is the amount of irregular current variation when the contact arm is in motion.

As shown in Fig. 3-12, a constant direct current is applied through the potentiometer by means of an external source. A battery is the best source since it is free of any noise or ripple. An output voltage from the potentiometer is applied to the oscilloscope vertical channel. The internal recurrent sweep can be used provided that the sweep frequency is above 100 Hz. If the potentiometer is "quiet" there will be a straight horizontal trace with no vertical deflection. Any vertical deflection indicates noise.

1. Connect the equipment as shown in Fig. 3-12.
2. Place the oscilloscope in operation as described in the instruction manual. Switch on the internal recurrent sweep and set the sync selector to external so that the sweep will not be triggered by noise.

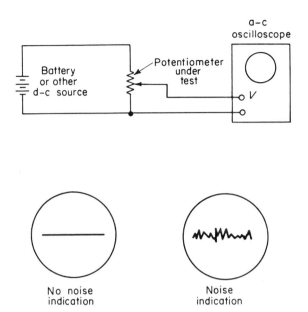

**Fig. 3-12.** Testing potentiometers for dynamic and static noise with an oscilloscope.

## NOTE

An a-c oscilloscope is recommended for this test since the voltage divider action of the potentiometer would move the trace vertically on a d-c oscilloscope.

3. Measure the static noise level, if any, on the voltage-calibrated scale. It is possible that a noise indication could be caused by pickup in the lead wires. If in doubt, disconnect the leads from the potentiometer but not from the oscilloscope. If the noise is still there, it is pickup noise; if the noise is gone, it is static noise (probably due to poor contact of the potentiometer arm).

4. Vary the potentiometer contact arm from one extreme to the other. Measure the dynamic noise level, if any, on the voltage-calibrated scale. Dynamic noise should not be difficult to distinguish since it occurs only when the contact arm is in motion. (On commercial test units, the contact arm is driven by a motor.)

## NOTE

The dynamic noise level will usually be increased if the battery voltage (or other source) is increased. Do not exceed the maximum rated voltage of the potentiometer when making these tests.

## 3-9. Measuring Thermal or Ballast Resistances

Some resistors are made of materials that increase in resistance when the temperature increases. On the other hand some resistors decrease in resistance value with temperature increases. All resistors are subject to variation between "hot" and "cold" resistance values. However, some resistors are manufactured specifically to produce a large change in resistance for change in temperature. The "Globar" resistor used in many radio receivers is an example of a temperature-sensitive resistor. The cold resistance is many times (as much as 10 to 200) that of the hot resistance. Once heated, the Globar resistor will remain in the low resistance condition for several minutes (or longer). The most practical way to test such a component is to heat the resistor, either by operation in the circuit or by placing a voltage across the resistor temporarily and then measuring the hot resistance. The cold resistance can be measured before heating the resistor and the hot-versus-cold resistance values compared.

Some resistors and resistance elements do not retain their "hot" resistance and must be measured while operating in a circuit. A vacuum tube filament (or heater) is a good example. The usual tube heater will be less than 5 ohms cold, and as high as 50 ohms hot.

The in-circuit value of a resistor can be found by measuring the current and voltage and then calculating resistance using Ohm's law, $R = E/I$. The basic test circuit is shown in 3-13. Once the hot resistance is found, the resistor can be removed from the circuit, and the cold resistance measured with an ohmmeter.

A *ballast* resistor can also be checked using the same test circuit (Fig. 3-13). Most ballast resistors have a positive temperature coefficient (resistance increases with temperature) to maintain constant current flow, even with changes in voltage. If voltage increases in a normal circuit, current

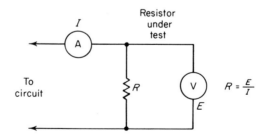

**Fig. 3-13.** Measuring in-circuit "hot" resistance.

flow will also increase. Under these conditions, a series-connected ballast resistor will become hotter, its resistance will increase, and the current flow will be lowered. Also, the voltage drop across the ballast resistor will be greater, lowering the voltage to the load, and the voltage and current will be maintained constant.

In practice, there will always be some increase in current through a ballast resistor with an increase in voltage. The amount can be checked using the test circuit of Fig. 3-13 and a variable power source. Increase the voltage until the current appears to level off; then increase the voltage in small steps and note the change in current for each step. Be careful not to exceed the maximum rated voltage and current for the ballast resistor.

## 3-10. Measuring Internal Resistance of Circuits

It is sometimes convenient to measure the internal resistance of a circuit while the circuit is operating. For example, it may be desired to measure the plate resistance of a tube or the collector resistance of a transistor. Obviously, the circuit resistance cannot be measured with an ohmmeter while the circuit is energized. However, it is possible to measure circuit resistance using a voltmeter and potentiometer.

### 3-10.1. Basic Method

1. Connect the equipment as shown in Fig. 3-14.

2. Set the potentiometer to zero and measure the circuit voltage. This will be the full circuit voltage.

3. Increase the potentiometer resistance until the voltage is one-half the full circuit voltage (one-half that obtained in Step 2).

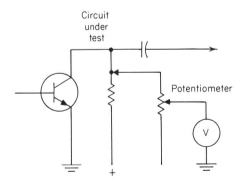

**Fig. 3-14.** Measuring internal resistance of circuits.

4. Disconnect the potentiometer from the circuit. Measure the d-c resistance of the potentiometer.

5. Subtract the input resistance of the meter from the d-c resistance of the potentiometer. The remainder is equal to the internal resistance of the circuit.

For example, assume that the meter reads 100 V when the potentiometer is set to zero and 50 V with the potentiometer set to 100,000 ohms. Also assume that the input resistance of the meter is 30,000 ohms. The circuit internal resistance is 70,000 ohms.

### NOTE

The input resistance of a meter should be listed in the manufacturer's data. If not, it is possible to calculate the input resistance using the circuit of Fig. 3-15. First, set the meter to the voltage scale or range that is to be used (a VOM usually shows a different input resistance for each voltage range). Connect the meter as shown in Fig. 3-15a. Set the voltage source to some convenient reading on the meter. Connect potentiometer $R$ into the circuit as shown in Fig. 3-15b. Increase the resistance of $R$ until the voltage is one-half of that previously obtained (without $R$ in the circuit). Disconnect $R$ from the circuit. Measure the d-c resistance of $R$. This value is equal to the input resistance of the meter.

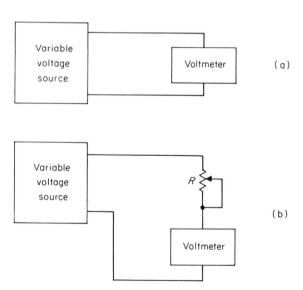

**Fig. 3-15.** Measuring input resistance of a voltmeter.

### 3-10.2. Power Supply Internal Resistance

Inexperienced technicians often assume that the internal resistance of a power supply can be found by dividing the output voltage by the current. However, this figure is the *load resistance,* not the power supply internal resistance.

Power supply resistance is determined by

$$\frac{\text{no-load voltage} - \text{full-load voltage}}{\text{current (amperes)}}$$

For example, if the no-load voltage is 100, the full-load voltage is 90, and the current is 500 mA.

$$\frac{100 - 90}{.5} = 20 \text{ ohms}$$

A low internal resistance is the most desirable since it indicates that the output voltage will change very little with load.

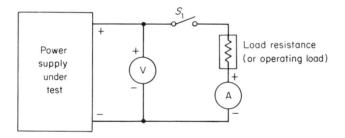

Fig. 3-16. Measuring power supply internal resistance.

1. Connect the equipment as shown in Fig. 3-16. Use a load resistance that will be equal to the operating load (or where practical, use the operating load).

2. Measure the power supply voltage without the load.

3. Apply the load and measure the power supply voltage again.

4. Measure the load current.

5. Using the equation, find the internal resistance.

# CHAPTER 4

# Capacitance Measurements

## 4-1. Basic Capacitance Equations

Although capacitors can be used in a variety of circuit combinations, these circuits are a version of the basic series, parallel, or series-parallel arrangements shown in Figs. 4-1 through 4-4.

The relationship of capacitive reactance and impedance of capacitors, together with the calculations, is shown in Fig. 4-5.

Unlike voltages across resistors, it is not possible simply to add the a-c voltages across the capacitor and resistor in a series circuit to find the source voltage. Instead, the individual voltages must be added vectorially. Figure 4-6 shows the relationship of voltages, together with the calculations, for voltages in series RC circuits.

When capacitors are used as *voltage dividers,* the source voltage will be distributed among the capacitors in inverse proportion to their capacitance (lowest voltage across largest capacitance). This relationship is shown in Fig. 4-7.

## 4-2. Capacitor Color Codes

Both ceramic and mica capacitors are color coded as to capacitance value, tolerance, temperature coefficient, etc. The color-coding system for modern mica capacitors is shown in Fig. 4-8. Note that the lower left-hand

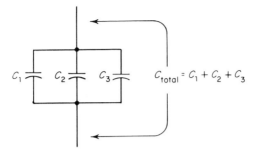

Any number of capacitors can be added in this
manner, provided the same capacitance units
(farad, microfarad, picofarad) are used.

**Fig. 4-1.** Calculations for total capacitance of capacitors in parallel.

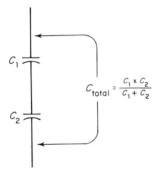

$C_{total}$ is always less than the value of either
capacitor. If $C_1 = C_2$, then $C_{total}$ is one-half of
$C_1$ or $C_2$. Where there is a 10 to 1 (or better)
ratio between $C_1$ and $C_2$, $C_{total}$ will be slightly
less than the smallest capacitor value.

**Fig. 4-2.** Calculations for total capacitance of two capacitors in series only.

"type" color dot indicates the type or classification of the particular capacitor according to the manufacturer's specifications. Such specifications may
include temperature coefficient, Q factor, and related characteristics. The
type designation is not necessarily the same for all manufacturers.

The color-coding system for modern ceramic capacitors is shown in
Fig. 4-9. As shown, a temperature-coefficient color band or dot is included along with the capacitance and tolerance color coding.

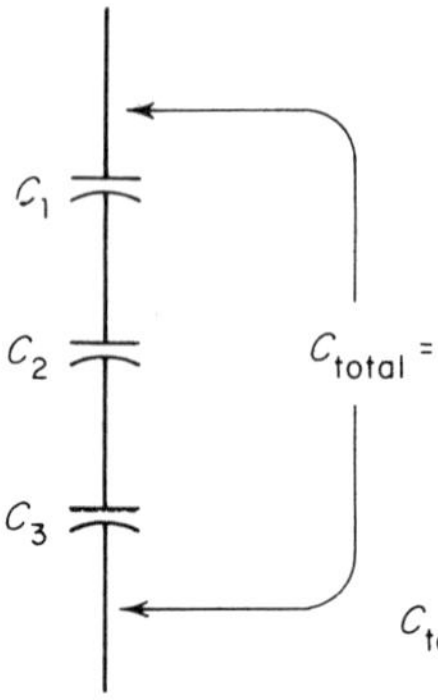

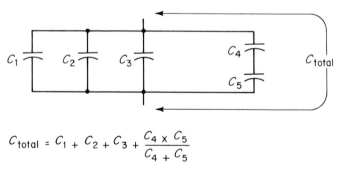

$$C_{total} = \cfrac{1}{\dfrac{1}{C_1} + \dfrac{1}{C_2} + \dfrac{1}{C_3} + \cdots}$$

Alternate

$$C_{total} = \cfrac{C_1}{1 + \dfrac{C_1}{C_2} + \dfrac{C_1}{C_3} + \dfrac{C_1}{C_4} + \cdots}$$

Alternate

$$\frac{1}{C_{total}} = \frac{1}{C_1} + \frac{1}{C_2} + \frac{1}{C_3} + \cdots$$

For convenience, make $C_1$ the largest capacitor in the series combination.

**Fig. 4-3.** Calculations for total capacitance of three (or more) capacitors in series.

$$C_{total} = C_1 + C_2 + C_3 + \frac{C_4 \times C_5}{C_4 + C_5}$$

**Fig. 4-4.** Calculations for total capacitance of series-parallel capacitor combinations.

The temperature coefficient of ceramic capacitors is given in parts per million per degree centigrade (ppm/°C). A preceding letter, N, denotes negative temperature coefficient (capacity decrease with an increase in operating temperature). The P designation indicates a positive temperature coefficient, and NPO indicates a negative-positive-zero type.

An N220 designation indicates a capacity decrease with a temperature rise of 220 ppm/°C, and shows how much the value changes during the

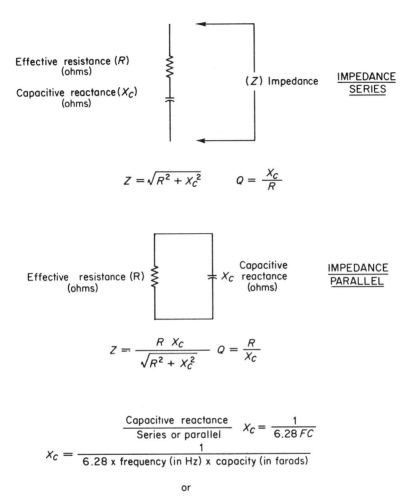

Effective resistance ($R$)
(ohms)

Capacitive reactance ($X_C$)
(ohms)

($Z$) Impedance

IMPEDANCE
SERIES

$$Z = \sqrt{R^2 + X_C^2} \qquad Q = \frac{X_C}{R}$$

Effective resistance (R)
(ohms)

$X_C$ Capacitive reactance
(ohms)

IMPEDANCE
PARALLEL

$$Z = \frac{R \; X_C}{\sqrt{R^2 + X_C^2}} \qquad Q = \frac{R}{X_C}$$

Capacitive reactance
Series or parallel

$$X_C = \frac{1}{6.28 \, FC}$$

$$X_C = \frac{1}{6.28 \times \text{frequency (in Hz)} \times \text{capacity (in farads)}}$$

or

$$X_C = \frac{159.2}{\text{frequency (in KHz)} \times \text{capacity (in microfarads)}}$$

$$C = \frac{1}{6.28 F X_C} \qquad F = \frac{1}{6.28 \, C X_C}$$

Fig. 4-5. Calculations for capacitive reactance and impedance in resistance-capacitance circuits.

warm-up time of the device in which the capacitor is used. The NPO types are stable units with negligible temperature effect on capacitance.

Where a value in Fig. 4-9 is marked *alternate,* this indicates a coding that may have been used in the past, but modern components are generally coded by the *preferred* coding designation.

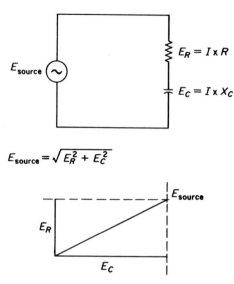

**Fig. 4-6.** Adding alternating current voltages in a series resistance-capacitance circuit.

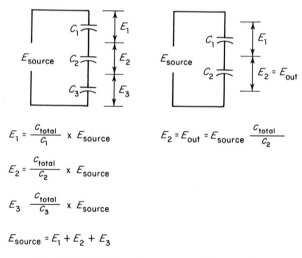

**Fig. 4-7.** Calculating voltages across series capacitors.

## 4-3. Capacitance Bridges

Precision capacitance values are often measured by means of bridge circuits. Although there are many capacitance bridge circuits in use, the resistance-ratio (Fig. 4-10), Wien (Fig. 4-11), and Schering (Fig. 4-12) are the best known. Operation of these circuits is similar to that of the resistance bridges described in Section 3-7.

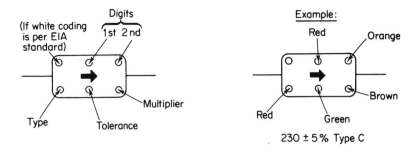

Capacitance Values in Picofarads (pF)

| Color | Digit | Multiplier | Tolerance | Type classification |
|---|---|---|---|---|
| Black | 0 | 1 | 20 % (±) | A |
| Brown | 1 | 10 | 1 % | B |
| Red | 2 | 100 | 2 % | C |
| Orange | 3 | 1000 | 3 % | D |
| Yellow | 4 | 10,000 | — | E |
| Green | 5 | — | 5 % | — |
| Blue | 6 | — | — | — |
| Violet | 7 | — | — | — |
| Gray | 8 | — | — | — |
| White | 9 | — | 10 % | — |
| Silver | — | 0.01 | — | — |
| Gold | — | 0.1 | — | — |

**Fig. 4-8.** Mica capacitor color codes.

From a practical standpoint, commercial bridge units should be used instead of homemade circuits. When bridge circuits are constructed in the laboratory from basic components, stray electromagnetic and electrostatic fields can easily be large enough to upset all measurements. High quality commercial instruments use every known method to eliminate these effects. For example, grounded shields are placed around all components, the galvanometer may be coupled into the circuit by a shielded output transformer, etc. Since commercial bridges are provided with detailed operating instructions, such data will not be duplicated here.

## 4-4. Testing Capacitors with a Meter

The obvious test of a capacitor is to check for leakage with an ohmmeter. It is also possible to find the approximate value of a capacitor with an ohmmeter. It is also possible to check operation of a capacitor under

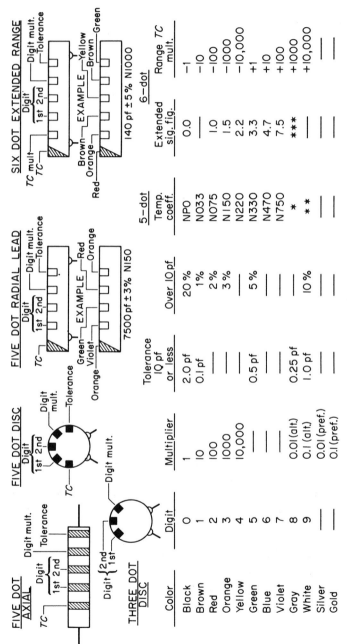

| Color | Digit | Multiplier | Tolerance 10 pf or less | Over 10 pf | 5-dot Temp. coeff. | Extended sig. fig. | Range TC mult. |
|-------|-------|-----------|------------------------|-----------|-------------------|--------------------|----------------|
| Black | 0 | 1 | 2.0 pf | 20% | NPO | 0.0 | −1 |
| Brown | 1 | 10 | 0.1 pf | 1% | N033 | 1.0 | −10 |
| Red | 2 | 100 | — | 2% | N075 | 1.5 | −100 |
| Orange | 3 | 1000 | — | 3% | N150 | 2.2 | −1000 |
| Yellow | 4 | 10,000 | — | — | N220 | 3.3 | −10,000 |
| Green | 5 | — | 0.5 pf | 5% | N330 | 4.7 | +1 |
| Blue | 6 | — | — | — | N470 | 7.5 | +10 |
| Violet | 7 | — | — | — | N750 | 7.5 | +100 |
| Gray | 8 | 0.01 (alt.) | 0.25 pf | — | * | *** | +1000 |
| White | 9 | 0.1 (alt.) | 1.0 pf | 10% | ** | — | +10,000 |
| Silver | — | 0.01 (pref.) | — | — | — | — | — |
| Gold | — | 0.1 (pref.) | — | — | — | — | — |

Capacitance values in picofarads (pf)

*General-purpose types with a temperature coefficient ranging from P150 to N1500.
**Coupling, decoupling and general bypass types with a temperature coefficient ranging from P100 to N750.
***If the first band (temperature coefficient) is black, the range is N1000 to N5000.

**Fig. 4-9.** Ceramic capacitor color codes.

68

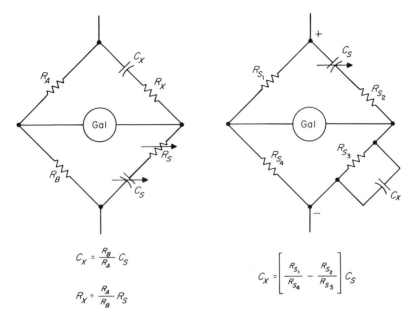

$$C_X = \frac{R_B}{R_A} C_S$$

$$R_X = \frac{R_A}{R_B} R_S$$

**Fig. 4-10.** Basic resistance-ratio bridge for capacitance measurement.

$$C_X = \left[ \frac{R_{S_1}}{R_{S_4}} - \frac{R_{S_2}}{R_{S_3}} \right] C_S$$

**Fig. 4-11.** Basic Wien bridge for capacitance measurement.

various conditions with a voltmeter. The following paragraphs describe these procedures.

### 4-4.1. Checking Capacitor Leakage with an Ohmmeter

The capacitor must be disconnected from the circuit to make an accurate leakage test. The basic leakage test is similar to measuring any high resistance value. The ohmmeter is connected across the capacitor terminals, and the resistance is measured. The following notes should be observed.

1. Use the highest resistance range of the ohmmeter. A typical capacitor will have a resistance in excess of 1000 megohms.

2. A more accurate test can be made if a high-ohms adapter (Section 3-6) is used. The higher voltage will show up any tendency of the capacitor to break down.

3. When using higher voltages, make certain not to exceed the voltage rating of the capacitor. This is a problem with the low-voltage electrolytic capacitors often used in solid-state equipment.

4. Another precaution to be observed when checking electrolytics is to make certain of the ohmmeter battery polarity. One ohmmeter terminal

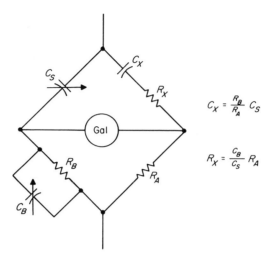

$$C_X = \frac{R_B}{R_A} C_S$$

$$R_X = \frac{C_B}{C_S} R_A$$

**Fig. 4-12.** Basic Schering bridge for capacitance measurement.

or lead will be positive, and the other lead negative. Usually, the positive terminal or lead is red, while the negative is black. However, to make sure, check the polarity against the meter schematic diagram or with an external voltmeter connected to the ohmmeter leads.

Usually, capacitors will indicate some measurable resistance when the ohmmeter leads are first connected, but then the indication will increase to infinity. This temporary resistance indication is caused by current flow as the capacitor charges. If the resistance indication remains below 1000 megohms after the capacitor is charged, it is likely that the capacitor is leaking. On the other hand, if the resistance indication remains at infinity (the pointer never moves when the ohmmeter leads are connected to the capacitor), it is possible that the capacitor is open. Either of these conditions should be followed up with a further test.

Sometimes it is possible to charge a capacitor faster if the ohmmeter leads are first connected with the lowest range ($R \times 1$) in use. Then select each higher ohmmeter range, in turn, until the highest range is in use. (The lowest ohmmeter range applies the most voltage.)

### 4-4.2. Checking Capacitor Leakage with a Voltmeter

If a capacitor is suspected of leakage or of being open, as indicated by improper circuit operation or the ohmmeter test just described, the facts can be confirmed using a voltmeter test.

If the capacitor is out-of-circuit, connect the capacitor leads across a

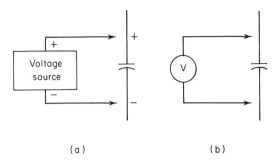

(a)                    (b)

**Fig. 4-13.** Testing capacitors for leakage with a voltmeter.

voltage source and hold for approximately 10 seconds. (See Fig. 4-13a.) For best results, use a voltage near the working voltage of the capacitor. Of course, never exceed the working voltage and always observe polarity for electrolytics.

With the capacitor charged, remove the voltage source and measure the *initial* capacitor voltage (with a voltmeter, *not an ohmmeter*). The initial voltage indication should be approximately the same as the source voltage. (See Fig. 4-13b.) If no voltage is indicated, the capacitor is open. If the voltage is very low, the capacitor is leaking.

This test may not be too effective in testing low-value capacitors, especially with a VOM. The low input resistance of a VOM could discharge a low-value capacitor too quickly to produce a measurable voltage indication. However, an electronic meter will discharge the capacitor slowly because of the high input resistance.

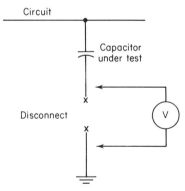

**Fig. 4-14.** Testing capacitors for leakage (while in-circuit).

If the capacitor is in-circuit, disconnect the capacitor ground lead, then measure d-c voltage from the lead to ground as shown in Fig. 4-14. Initially, there may be some d-c voltage indication due to capacitor charge. (The exact amount of initial indication will depend upon input resistance of the meter and capacitor value.) However, if the d-c voltage indication remains, the capacitor is leaking. It may be necessary to measure the initial charging indication on a high voltage range of the meter. If the voltage indication drops, move the meter to the lowest voltage range to measure possible small voltages due to high resistance leakage of the capacitor.

If the capacitor shows no voltage indication from the ground lead to ground, the capacitor is definitely not leaking. However, there is still the possibility of an open capacitor. A large-value capacitor will show an indication in most circuits, but as in the case of the resistance test, a low-value capacitor may charge too quickly to produce a measurable reading.

### 4-4.3. Checking Capacitors by Signal Tracing

An in-circuit open capacitor can be checked quickly and positively using a voltmeter equipped with a signal-tracing probe. Of course, there must be a signal present in the circuit. If necessary, connect a signal generator to the input of the circuit.

Figure 4-15a shows the basic circuit for checking a coupling capacitor, including a signal-tracing probe. Under usual conditions, the a-c voltage (or signal) should be the same on both the input and output sides of a coupling capacitor. There may be some attenuation of the voltage on the output side (output will be lower than input). The complete absence of a signal at the output side of the capacitor indicates an open capacitor (or excessive leakage).

Figure 4-15b shows the basic circuit for checking a bypass capacitor (screen, cathode, emitter, etc.). Under usual conditions, the a-c voltage (or signal) *across* a bypass capacitor will be large if the capacitor is open.

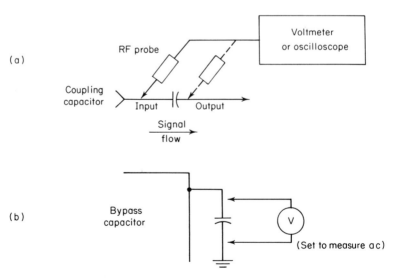

**Fig. 4-15. Testing capacitors by signal tracing.**

The usual function of a bypass capacitor is to pass a-c voltage or signals to ground. Therefore, there should be no a-c voltage across the capacitor.

### 4-4.4. Measuring Capacitor Values with a Voltmeter

It is possible to find the *approximate* value of a capacitor using a voltmeter. The method is based on the time constants of an RC circuit. A capacitor of a given value will charge to 63.2% of its full value in a given number of seconds through a given resistance value. Likewise, a capacitor will discharge to 36.8% of its full value through a given resistance in a given time. If the resistance value is known and the time measured, the capacitance value can be calculated.

The basic discharge-method circuit is shown in Fig. 4-16, with the charge-method circuit shown in Fig. 4-17.

Either a VOM or electronic voltmeter can be used. The input resistance of the meter must be known. The input resistance of a VOM is equal to the ohms/V rating multiplied by the full-scale voltage. Therefore, the VOM input resistance will change with each voltage range selected. The input resistance of an electronic voltmeter remains constant and usually ranges from 11 to 16 megohms. Consult the meter manufacturer's data for the input resistance.

The high input resistance of an electronic voltmeter will cause the capacitor to charge and discharge slowly. Therefore, the electronic voltmeter should be used with small value capacitors.

Use the following procedure with the *discharge circuit* of Fig. 4-16.

1. Move switch $S_1$ to the charge position and hold for approximately one minute.

2. Move switch $S_1$ to the discharge (meter) position. Simultaneously, start a stopwatch or note the exact time on the second hand of a conventional watch.

3. When the voltage has dropped to 36.8% of its initial value, stop the watch or note the elasped time interval. For convenience, use an even voltage value such as 100 V, 10 V, etc.

4. Divide the elapsed discharge time by the input resistance of the meter to find the capacitance in farads. If the input resistance of the meter is converted to megohms, the capacitance value will be in microfarads.

5. As an example, assume that the capacitor is charged to an initial value of 10 V, that the capacitor discharges to 3.68 V at the end of 90 seconds, and that the input resistance of the meter is 2 megohms (200,000 ohms/V on the 10 V full-scale range). Therefore $90/2 = 45\ \mu F$.

Use the following procedure with the *charge circuit* of Fig. 4-17.

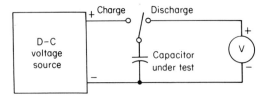

**Fig. 4-16.** Measuring capacitor values with a voltmeter (charge-discharge method).

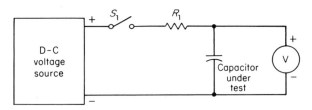

**Fig. 4-17.** Measuring capacitor values with a voltmeter (charge method).

1. Make certain that there is no voltage indicated across the capacitor from a previous charge.

2. Close switch $S_1$. Simultaneously, start a stopwatch or note the exact time on the second hand of a conventional watch.

3. When the voltage reaches 63.2% of its final value, stop the watch or note the elapsed time interval. For convenience, use an even voltage value such as 100 V, 10 V, etc. Also use an even $R_1$ resistance value, preferably in megohms so the capacitance value can be expressed in microfarads. Note that there will be some voltage drop across $R_1$. Therefore, the capacitor will never fully charge to the value of the source voltage. Instead, the capacitor charge will depend upon the ratio of $R_1$ resistance value to the internal resistance of the meter. For example, if $R_1$ is 1 megohm, the meter is 3 megohms, and the source is 100 megohms, the capacitor will not charge beyond about 75 V

$$1 \text{ megohm} + 3 \text{ megohms} = 4 \text{ megohms}$$

$$\frac{100}{4} = 25; \qquad 100 \text{ V} - 25 \text{ V} = 75 \text{ V}$$

4. Divide the elapsed charge time by the value of series resistance $R_1$ to find the capacitance in farads (if $R_1$ is in megohms, the capacitance value will be in microfarads).

5. For example, assume that the capacitor is charged to an initial value of 100 V (by an approximately 133-V source), that the capacitor charges to 63.2 V at the end of 70 seconds, and that the resistance of $R_1$ is 3 megohms. $70/3 = 23.3 \ \mu\text{F}$.

Remember that these tests will provide an *approximate* capacitance value at best and will not be accurate with a capacitor that is leaking. However, the test will show that the capacitor is operating normally and that its approximate value is correct.

## 4-5. Measuring Capacitance with a Dip Circuit

The value of a capacitor can be found using a dip circuit. These circuits use the same principle as is found in commercial dip meters, except that a signal generator is used as the signal source. The accuracy of the signal generator determines the accuracy of the circuit.

In practice, a dip circuit measures the resonant frequency of a tuned LC circuit. There is no direct connection to the external circuit, and the circuit need not be energized. (Most dip circuit measurements are made with the external circuit "cold.")

A basic resonant circuit consists of a capacitance and an inductance. Either of these two values can be determined by substitution if the other is known and the resonant frequency is known.

Figure 4-18 shows how the basic dip circuit operates. The dip circuit pickup coil $L_1$ is coupled to an external resonant circuit. The signal generator is adjusted to produce an unmodulated RF output. This output is rectified by diode $CR_1$ and appears as a d-c reading on microammeter $M_1$. When the signal generator is tuned to the resonant frequency of the circuit or vice versa, part of the energy from coil $L_1$ is absorbed into the resonant circuit. This causes the meter $M_1$ indication to drop or "dip." The resonant frequency of the signal generator is then equal to the resonant frequency of the external circuit.

When a dip circuit is used to measure the value of an unknown capacitor, the capacitor is connected in parallel with an inductance of known

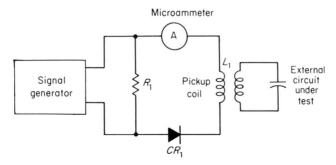

**Fig. 4-18. Basic dip adapter circuit.**

value. This resonant circuit is then coupled to the basic dip circuit of Fig. 4-18, and the resonant frequency is found by adjusting the signal generator for a dip on the microammeter. With the resonant frequency and the inductance value known, the capacitance value can be calculated using this equation

$$\frac{\text{capacitance}}{(\text{picofarads})} = \frac{2.53 \times 10^4}{(\text{frequency in megahertz})^2 \times (\text{inductance in microhenrys})}$$

### 4-5.1. Basic Procedure

1. Connect the equipment as shown in Fig. 4-19. If the test is to be repeated often, the test inductance leads should be provided with alligator clips. This will permit easy connection to the capacitor under test. In any event, use the shortest possible leads between the known inductance and the unknown capacitance.

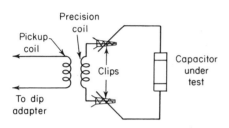

Fig. 4-19. Basic capacitance measurement with dip adapter.

2. Use some convenient value of inductance, such as 100, 200, or 300 $\mu$H. Both the inductance and *distributed capacitance* of the test coil must be known. The actual inductance and distributed capacitance should be measured; the manufacturer's data should not be relied on. The procedure for determining inductance and distributed capacitance is described in Chapter 5. The best results will be obtained when a high-Q inductance is used.

3. Set the generator to a frequency well below that of the assumed resonant frequency. If the approximate resonant frequency is not known, set the generator to its lowest frequency. Adjust the generator output for a good mid-scale indication on the dip circuit meter.

4. Slowly increase the generator frequency, watching for a dip indication on the meter. Adjust the generator to that frequency that produces the bottom of the dip indication.

### NOTE

The nature of the dip indication will provide an *approximate* reflection of the capacitor Q or quality. A broad dip indicates a low-Q, while a sharp dip indicates a high-Q. Of course, the quality of the test circuit and inductance will also have some effect on the dip indica-

tion. However, the dip indication can be used as a basis for comparison of similar capacitors.

5. Once the resonant frequency has been found, calculate the capacitor
value using the equation. For example, assume a resonant frequency of
3 MHz and a test inductance of 100 $\mu$H.

$$C = \frac{2.53 \times 10^4}{3^2 \times 100} = 28.1 \text{ pF}$$

### NOTE

If the amount of distributed capacitance in the test coil is small in
relation to the unknown capacitance, the distributed capacitance can
be ignored. However, in the example given the distributed capacitance of a typical test inductance would probably have some effect on
the measurement. Therefore, the distributed capacitance must be
known and *subtracted* from the calculated capacitance (since both
capacitance values are in parallel and are additive). Thirty pF is about
the lowest value of capacitance that can be measured with the basic
dip circuit. This can be extended to about 10 pF if the dip circuit is
modified with a parallel fixed capacitor as shown in Fig. 4-20. The
actual value of the known capacitor is not of particular importance,
but it must be known and it must be accurate. For example, assume
that a 100-pF parallel capacitor is used to measure a 10-pF capacitor.
If the 100-pF test capacitor has a 1% tolerance (1 pF in 100 pF)
this would produce a 10% inaccuracy (1 pF in 10 pF).

6. Once the parallel test capacitor has been connected, the procedure
for finding the unknown capacitance is the same as for the basic dip circuit.
Once the resonant frequency has been found, calculate the capacitor value
using the equation. Then *subtract* the value of the known parallel capacitor. For example, assume that the parallel capacitor had a value of 100

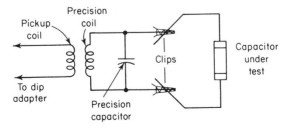

**Fig. 4-20.** Capacitance measurement with parallel fixed capacitor.

pF, with a 2% or 2-pF tolerance, and the resonant dip calculation shows a value of 113 pF, the unknown capacitor value would be 11 to 15 pF (113 − 100 = 13 ±2 pF).

### 4-5.2. In-circuit Capacitance Measurement

The basic dip circuit procedure can be applied to an in-circuit capacitor. However, certain precautions must be observed. The in-circuit capacitance measurement is made in the same manner as the out-of-circuit test. Except for a parallel capacitor, other circuit elements can be ignored. If the capacitor being tested is in parallel with another capacitor (or any circuit element with measurable capacitance, such as a transistor or diode), a dip indication may be obtained from the other circuit element, even though the capacitor under test is defective.

In general, a sharp, well-defined dip indicates that the capacitor is good. If the dip is very broad (much broader than another capacitor of similar value and type), the capacitor is developing some leakage. If no dip is indicated, the capacitor is probably defective (open, shorted, or leaking badly).

### NOTE

The dip circuit method (both in-circuit and out-of-circuit) is not too effective for capacitors of very large or very small value, or with electrolytic capacitors.

## 4-6. Measuring Capacitance with a Resonant Circuit

The value of a capacitor can also be found using a resonant circuit, signal generator, and voltmeter. The procedure is similar to that used in the dip circuit. A resonant LC circuit is formed with an unknown capacitor and a known value of inductance. The resonant circuit is connected to the signal generator and meter as shown in Fig. 4-21. The resonant frequency of the circuit is found by tuning the signal generator for a peak or maximum indication on the meter. With the resonant frequency and the inductance value known, the capacitance value can be calculated.

### 4-6.1. Basic Procedure

1. Connect the equipment as shown in Fig. 4-21. The generator must be capable of producing a signal at the resonant frequency of the circuit, and the meter must be capable of measuring the frequency. If the resonant

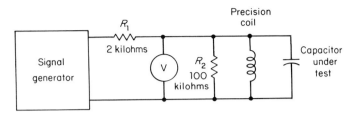

**Fig. 4-21.** Capacitance measurement with resonant circuit.

frequency is beyond the normal range of the meter, an RF probe must be used.

2. Use some convenient value of inductance, such as 100, 200, or 300 $\mu$H. Measure the actual inductance and distributed capacitance as described in Chapter 5.

3. Adjust the generator output until a convenient mid-scale indication is obtained on the meter. Use an unmodulated signal output from the generator.

4. Starting at a frequency well below the lowest possible resonant frequency of the inductance-capacitance combination under test, slowly increase the generator frequency. If there is no way to judge the approximate resonant frequency, use the lowest generator frequency.

5. Watch the meter for a maximum or peak indication. Note the generator frequency at which the peak indication occurs. This is the resonant frequency of the circuit.

### NOTE

The nature of the peak indication will provide an approximate reflection of the capacitor Q or quality. A broad peak indicates a low-Q, while a sharp peak indicates a high-Q.

6. Using the resonant frequency, and the known inductance value, calculate the unknown capacitance using the equation

$$\frac{\text{capacitance}}{(\text{picofarads})} = \frac{2.53 \times 10^4}{(\text{frequency in megahertz})^2 \times (\text{inductance in microhenrys})}$$

### NOTE

If the amount of distributed capacitance in the test coil is small in relation to the unknown capacitance, the distributed capacitance can be ignored. Otherwise, the distributed capacitance must be subtracted from the calculated capacitance.

# CHAPTER 5

# Inductance Measurements

## 5-1. Basic Inductance Equations

Inductors (coils) can be connected in series, parallel, and in series-parallel. Where no interaction of magnetic fields is produced by the inductors, the equations are as shown in Figs. 5-1 through 5-4.

Unless inductors (coils) are completely surrounded by shielding, are placed at right angles to each other, or are sufficiently far apart, the magnetic field of one inductance will affect the other inductance. This effect is known as *mutual inductance* and is measured in henrys (or millihenrys or microhenrys). Mutual inductance depends upon the *self-inductance* of the inductors and upon the closeness (or efficiency) with which the inductors are coupled. The effectiveness of the mutual coupling is known as the *coupling coefficient* or *k* factor. A 100% coupling (impossible in practical applications) would be a *k* factor of 1; 90% efficiency would be a *k* factor of 0.9. The calculations for mutual inductance and coupling coefficient are shown in Fig. 5-5.

Mutual inductance may add to the self-inductance of inductors or may subtract from the self-inductance, depending upon how the inductors are arranged. If the current flows in the same direction through each coil, the magnetic fields reinforce each other, and the mutual coupling increases. If the current flows in opposite directions, the mutual coupling decreases. The calculations for the effects of mutual inductance are shown in Fig. 5-6.

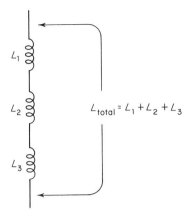

$$L_{total} = L_1 + L_2 + L_3$$

Any number of inductors can be added in this
manner, provided the same inductance units (henry,
millihenry, microhenry) are used. $L_{total}$ is correct
when there is no magnetic interaction.

**Fig. 5-1.** Calculations for total inductance of coils in series.

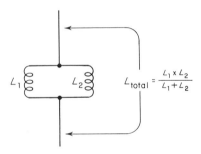

$$L_{total} = \frac{L_1 \times L_2}{L_1 + L_2}$$

$L_{total}$ is correct when there is no magnetic interaction.
$L_{total}$ is always less than the value of either coil. If $L_1 = L_2$,
then $L_{total}$ is one-half $L_1$ or $L_2$. When there is a 10 to 1
(or better) ratio between $L_1$ and $L_2$, $L_{total}$ will be
slightly less than smallest coil value.

**Fig. 5-2.** Calculations for total inductance of two coils in parallel only.

The relationship of inductive reactance and impedance of inductors,
together with the calculations, is shown in Fig. 5-7.

Unlike d-c voltages across resistors, it is not possible simply to add the
a-c voltages across the inductor and resistor in a series circuit to find the
source voltage. Instead, the individual voltages must be added vectorially.

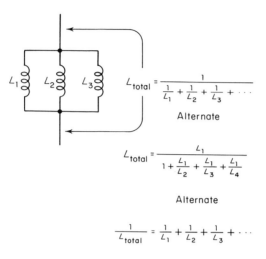

$$L_{total} = \cfrac{1}{\cfrac{1}{L_1} + \cfrac{1}{L_2} + \cfrac{1}{L_3} + \cdots}$$

Alternate

$$L_{total} = \cfrac{L_1}{1 + \cfrac{L_1}{L_2} + \cfrac{L_1}{L_3} + \cfrac{L_1}{L_4}}$$

Alternate

$$\frac{1}{L_{total}} = \frac{1}{L_1} + \frac{1}{L_2} + \frac{1}{L_3} + \cdots$$

$L_{total}$ is correct when there is no magnetic interaction. For convenience, make $L_1$ the largest coil in the parallel combination.

**Fig. 5-3.** Calculations for total inductance of three (or more) coils in parallel.

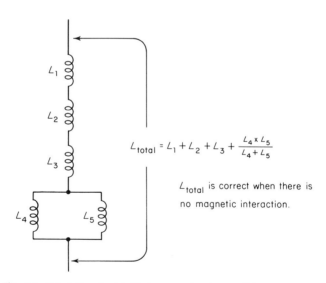

$$L_{total} = L_1 + L_2 + L_3 + \frac{L_4 \times L_5}{L_4 + L_5}$$

$L_{total}$ is correct when there is no magnetic interaction.

**Fig. 5-4.** Calculations for total inductance of series-parallel coil combinations.

Figure 5-8 shows the relationship of voltages, together with the calculations for voltages in series RL circuits.

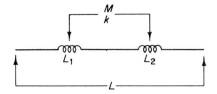

Mutual Inductance

$$M = k\sqrt{L_1 \times L_2}$$

Coupling Coefficient

$$k = \frac{M}{\sqrt{L_1 \times L_2}}$$

**Fig. 5-5.** Calculations for mutual inductance and coupling coefficient.

Inductors in Series Aiding

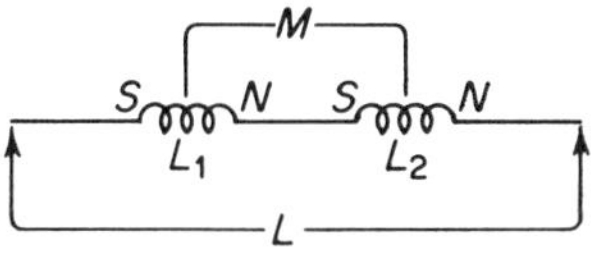

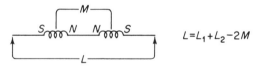

$$L = L_1 + L_2 + 2M$$

Inductors in Series Opposing

$$L = L_1 + L_2 - 2M$$

Inductors in Parallel Aiding

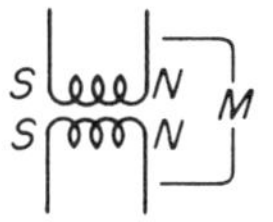

$$L = \frac{1}{\dfrac{1}{L_1+M} + \dfrac{1}{L_2+M}}$$

Inductors in Parallel Opposing

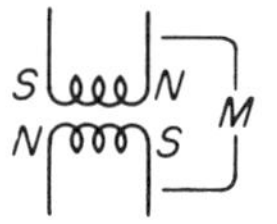

$$L = \frac{1}{\dfrac{1}{L_1-M} + \dfrac{1}{L_2-M}}$$

**Fig. 5-6.** Calculating the effects of mutual inductance.

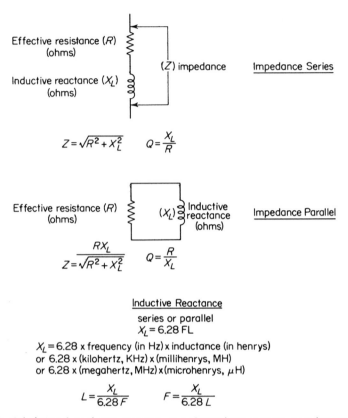

Effective resistance ($R$) (ohms)

Inductive reactance ($X_L$) (ohms)

($Z$) impedance      Impedance Series

$$Z = \sqrt{R^2 + X_L^2} \qquad Q = \frac{X_L}{R}$$

Effective resistance ($R$) (ohms)

($X_L$) Inductive reactance (ohms)      Impedance Parallel

$$Z = \frac{RX_L}{\sqrt{R^2 + X_L^2}} \qquad Q = \frac{R}{X_L}$$

Inductive Reactance
series or parallel
$X_L = 6.28\,FL$

$X_L = 6.28 \times$ frequency (in Hz) $\times$ inductance (in henrys)
or $6.28 \times$ (kilohertz, KHz) $\times$ (millihenrys, MH)
or $6.28 \times$ (megahertz, MHz) $\times$ (microhenrys, $\mu$H)

$$L = \frac{X_L}{6.28\,F} \qquad F = \frac{X_L}{6.28\,L}$$

**Fig. 5-7.** Calculations for inductive reactance, Q, and impedance in resistance-inductance circuits.

The equation necessary to calculate the self inductance of a single air-core coil is given in Fig. 5-9.

## 5-2. Inductance Bridges

Precision inductance values are often measured by means of bridge circuits. Although there are many basic inductance bridge circuits in use, the resistance-ratio (Fig. 5-10), Maxwell (Fig. 5-11), Hay (Fig. 5-12) and Owen (Fig. 5-13) bridges are the best known. Bridge circuits for measurement of mutual inductance include the Felici (Fig. 5-14), modified Campbell (Fig. 5-15), and the Carey-Foster (Fig. 5-16). Operation of these circuits is similar to that of the resistance bridges described in Section 3-7. The reasons for using commercial bridge circuits, together with their related instruction manuals, are discussed in Section 4-3.

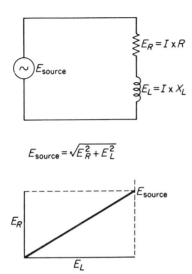

$$E_{source} = \sqrt{E_R^2 + E_L^2}$$

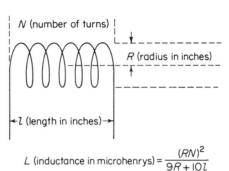

$E_R$

$E_L$

$L$ (inductance in microhenrys) $= \dfrac{(RN)^2}{9R + 10l}$

**Fig. 5-8.** Adding a-c voltages in a series resistance-inductance circuit.

**Fig. 5-9.** Calculations for self-inductance in a single air-core coil.

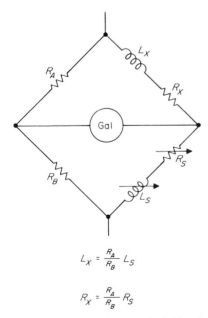

$$L_X = \frac{R_A}{R_B} L_S$$

$$R_X = \frac{R_A}{R_B} R_S$$

**Fig. 5-10.** Basic resistance-ratio bridge for inductance measurement.

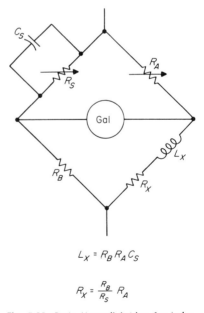

$$L_X = R_B R_A C_S$$

$$R_X = \frac{R_B}{R_S} R_A$$

**Fig. 5-11.** Basic Maxwell bridge for inductance measurement.

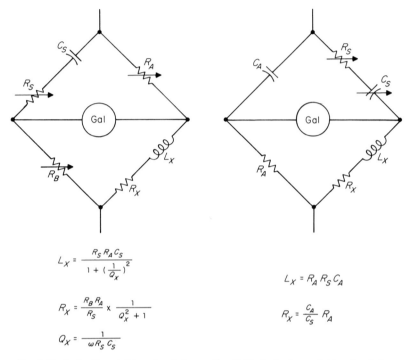

$$L_x = \frac{R_S R_A C_S}{1 + (\frac{1}{Q_x})^2}$$

$$R_x = \frac{R_B R_A}{R_S} \times \frac{1}{Q_x^2 + 1}$$

$$Q_x = \frac{1}{\omega R_S C_S}$$

$$L_x = R_A R_S C_A$$

$$R_x = \frac{C_A}{C_S} R_A$$

**Fig. 5-12.** Basic Hay bridge for inductance measurement.

**Fig. 5-13.** Basic Owen bridge for inductance measurement.

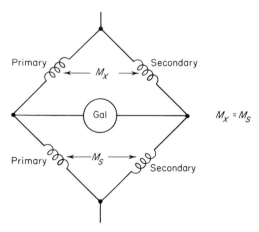

$$M_x = M_S$$

**Fig. 5-14.** Basic Felici bridge for mutual inductance measurement.

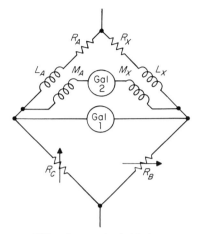

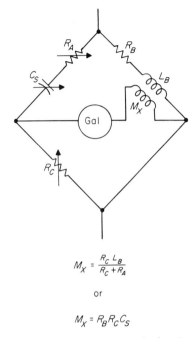

With galvanometer 1 at balance:

$$M_X = \frac{R_B}{R_C} \quad M_S = \frac{L_X}{L_A} \quad M_S.$$

With galvanometer 2 at balance:

$$L_X = \frac{R_B}{R_C} L_A \quad R_X = \frac{R_B}{R_C} R_A.$$

**Fig. 5-15.** Basic Campbell (modified) bridge for mutual inductance measurement.

$$M_X = \frac{R_C L_B}{R_C + R_A}$$

or

$$M_X = R_B R_C C_S$$

**Fig. 5-16.** Basic Carey-Foster bridge for mutual inductance measurement.

## 5-3. Measuring Inductance with a Dip Circuit

The value of an inductor can be found using a dip circuit. These circuits use the same principle as is found in commercial dip meters, except that a signal generator is used as the signal source. The accuracy of the signal generator determines the accuracy of the circuit.

In practice, a dip circuit measures the resonant frequency of a tuned LC circuit. There is no direct connection to the external circuit, and the circuit need not be energized. (Most dip circuit measurements are made with the external circuit "cold.")

A basic resonant circuit consists of a capacitance and an inductance. Either of these two values can be determined by substitution if the other is known and the resonant frequency is also known.

Operation of the basic dip circuit is shown in Fig. 4-18 and is described in Section 4-5. When the dip circuit is used to measure the value of an unknown inductance, the inductance is connected in parallel with a capacitor of known value. This resonant circuit is then coupled to the basic dip circuit of Fig. 4-18, and the resonant frequency is found by adjusting the signal generator for a dip on the microammeter. With the resonant frequency and the capacitance value known, the inductance value can be calculated using the equation

$$\frac{\text{inductance}}{\text{(microhenrys)}} = \frac{2.53 \times 10^4}{\text{(frequency in megahertz)}^2 \times \text{(capacitance in picofarads)}}$$

### 5-3.1. Basic Procedure

1. Connect the equipment as shown in Fig. 5-17. If the test is to be repeated often, the test capacitance leads should be provided with alligator clips. This will permit easy connection to the inductance under test. In any event, use the shortest possible leads between the known capacitance and the unknown inductance.

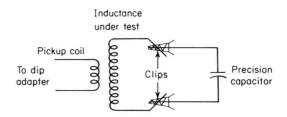

**Fig. 5-17.** Basic inductance measurement with a dip meter.

2. Use a convenient value of capacitance, such as 100, 200, or 300 pF. The best results will be obtained when a high-Q capacitor is used.

3. Set the generator to a frequency well below that of the assumed resonant frequency. If the approximate resonant frequency is not known, set the generator to its lowest frequency. Adjust the generator output for a good mid-scale indication on the dip circuit meter.

4. Slowly increase the generator frequency, watching for a dip indication on the meter. Adjust the generator to that frequency that produces the bottom of the dip indication.

### NOTE

The nature of the dip indication will provide an *approximate* reflection of the inductor Q. A broad dip indicates a low Q, while a sharp dip indicates a high Q. Of course, the quality of the test circuit and capacitance will also have some effect on the dip indication. However, the dip indication can be used as a basis for comparison of similar inductors.

5. Once the resonant frequency has been found, calculate the inductor value using the equation. For example, assume a resonant frequency of 7 MHz and a test capacitance of 100 pF.

$$L = \frac{2.53 \times 10^4}{7^2 \times 100} = 5.1 \ \mu\text{H}$$

## 5-4. Measuring Self-resonance and Distributed Capacitance of Inductors

There will be some distributed capacitance in any coil no matter what winding system is used. When the distributed capacitance combines with the coil's inductance, a resonant circuit is formed. This resonant frequency is usually quite high in relation to the frequency at which the coil will be used. However, since self-resonance may be at or near a harmonic of the frequency to be used, the self-resonant effect may limit the coil's usefulness in LC circuits. Some coils, particularly RF chokes, may have more than one self-resonant frequency.

### 5-4.1. Measuring Self-resonance

1. Remove the coil from its related circuit. This will eliminate the capacitance effect of nearby objects. If the coil cannot be removed from its circuit, keep the coil leads apart to minimize the capacitance effect.

2. Connect the dip circuit to the coil under test, using the most convenient method. Techniques for connecting (or coupling) dip circuits to inductors are described in Section 5-5. The direct capacitive-coupling method is not recommended when measuring self-resonance. Direct capacitive coupling produces a few pF of capacitance that are added to the coil-distributed capacitance and could result in an error.

3. If the dip circuit coil diameter is quite different from that of the coil being tested, the link coupling method is usually the most effective. If it is difficult to couple the link coil directly over the test coil, lay the dip circuit coil near the test coil with the turns parallel.

4. Set the generator to a frequency well below that of the assumed resonant frequency. If the approximate resonant frequency is not known, set the generator to its lowest frequency. Adjust the generator output for a good mid-scale indication on the dip circuit meter.

5. Slowly increase the generator frequency, watching for a dip indication on the meter. Adjust the generator to that frequency that produces the bottom of the dip indication. RF chokes will usually produce a broader indication than a simple air-core coil.

6. Since there may be more than one self-resonant point, tune through the entire signal generator range.

5-4.2. Measuring Distributed Capacitance

1. Once the resonant frequency has been found, calculate the distributed capacitance using the equation of Section 4-5.

2. For example, assume that a coil with an inductance of 7 $\mu$H was found to be self-resonant at 50 MHz.

$$C \text{ (distributed capacitance)} = \frac{2.53 \times 10^4}{50^2 \times 7} = 1.44 \text{ pF}$$

## 5-5. Coupling Dip Circuits to Inductors

When a basic dip circuit is coupled to an external resonant circuit, there must be some form of mutual coupling between the dip circuit coil and the resonant circuit. There are several methods for making such a coupling. The most common technique is to position the dip circuit coil near the external circuit and *parallel* to the external coil. This will produce simple inductive coupling, as shown in Fig. 5-18. The dip circuit coil can be positioned at the end of the external coil (Fig. 5-18a) or next to the external coil (Fig. 5-18b).

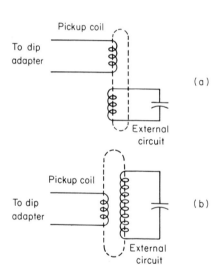

Fig. 5-18. Inductive coupling of dip adapter to external circuit.

When the measurements are made at ultrahigh frequencies, both the dip circuit coil and the external coil are usually "hairpin" loops. Such coils should be coupled as shown in Fig. 5-19.

Often the coil of the external circuit is mounted on a crowded chassis. Under these conditions it is nearly impossible to position the dip circuit coil close to the external circuit coil. Under these conditions, it is possible to link-couple the two coils as shown in Fig. 5-20. Three precautions must be observed when using link-coupling.

1. The amount of coupling is determined by the tightness of the link-coupling coil on the test-circuit coil, not by the length of the link-coupling.

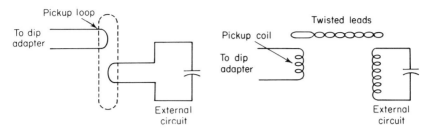

**Fig. 5-19.** Inductive coupling of UHF dip adapter to external UHF circuit.

**Fig. 5-20.** Link-coupling of dip adapter to external circuit.

2. The dip circuit end of the link-coupling should have a coil the same approximate diameter as the dip circuit coil.

3. The coupling will be inductive if the link-coupling coil is parallel to the coil being tested, and will be capacitive if the coupling coil is at right angles to the test coil.

When the dip circuit coil and test coil are parallel the coupling is inductive. It is possible to use capacitive coupling by positioning the two coils at right angles to each other as shown in Fig. 5-21. In general, capacitive coupling is used for high-Q circuits, while inductive coupling is used for low-Q circuits.

In extreme cases where the link-coupling will not reach the circuit to be tested, it is possible to use direct capacitive coupling, as shown in Fig. 5-22. This coupling method is not recommended since it will add a small amount of capacitance to the circuit being tested, thus causing the dip circuit to indicate a resonant frequency lower than the true value.

## 5-6. Dynamic Distributed Capacitance
### Measurements

Although the method of measuring distributed capacitance described in Section 5-4.2 is valid, it is often more convenient (and realistic) to measure

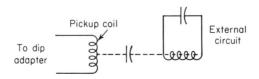

**Fig. 5-21.** Capacitive coupling of dip adapter to external circuit.

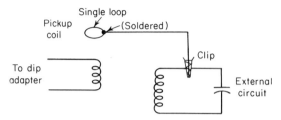

**Fig. 5-22.** Direct capacitive coupling of dip adapter to external circuit.

distributed capacitance while the coil is operating in its circuit. This dynamic measurement method requires a signal generator, meter, and two fixed capacitors with values of known accuracy. The signal generator must be capable of operation at the normal resonant frequency of the test circuit and at frequencies on either side of the resonant point. Likewise, the meter must be capable of measuring such frequencies. Often, an RF probe will be required for the meter. The actual values of the test capacitors are not critical.

1. Connect the equipment as shown in Fig. 5-23a.

2. Adjust the signal generator to the resonant frequency of the coil as indicated by a peak reading on the meter. Record this resonant frequency.

3. Connect one of the test capacitors across the coil and retune the signal generator for a resonant frequency indication (peak) on the meter. Record the second resonant frequency.

4. Connect the remaining test capacitor across the coil. Retune the signal generator for a resonant frequency indication (peak) on the meter. Record the third resonant frequency.

5. Plot the three points on a graph similar to that shown in Fig. 5-23b. The horizontal scale is used to mark frequency values in terms of $1/f^2$. For example, if the resonant frequency (with the coil alone) is 2 MHz, the frequency scale could be 0.25 for each division ($1/f^2 = 0.25$). The capacitance values are marked on the vertical scale, both above and below the center line. Actual graph spacing is not critical, but both the vertical and horizontal spaces must be equal. Actual capacitance and frequency values assigned to the spaces are not important, but the capacitance and frequency values must be realistic in relation to the anticipated distributed capacitance and must be consistent once assigned.

6. Draw a vertical line from the first point of resonant frequency (coil only, no test capacitors connected).

7. Draw a second line through the two frequency points with the capacitors connected. Extend this line until it intersects the vertical line.

Associated circuit

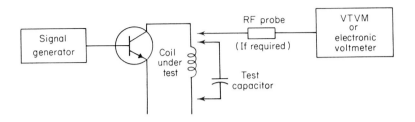

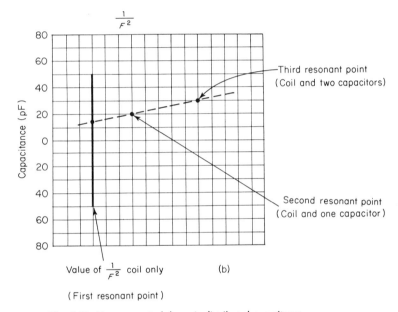

(b)

Value of $\frac{1}{F^2}$ coil only

(First resonant point)

**Fig. 5-23.** Measurement of dynamic distributed capacitance.

8. The point at which the two lines intersect represents the distributed capacitance of the coil and can be read off the vertical capacitance scale.

9. If the coil to be tested is connected in parallel with a capacitance, such as an LC tank circuit, the circuit capacitance must be subtracted from the value found in Step 8 to find the true distributed capacitance.

## 5-7. Measuring Inductance with a Resonant Circuit

The value of an inductance can also be found using a resonant circuit, signal generator, and voltmeter. The procedure is similar to that used in the dip circuit. A resonant LC circuit is formed with a coil of unknown inductance value and a capacitor of known value. The resonant circuit is connected to the signal generator and meter as shown in Fig. 5-24. The resonant frequency of the circuit is found by tuning the signal generator for a peak or maximum indication on the meter. With the resonant frequency and the capacitance value known, the inductance value can be calculated.

### 5-7.1. Basic Procedure

1. Connect the equipment as shown in Fig. 5-24. The generator must be capable of producing a signal at the resonant frequency of the circuit, and the meter must be capable of measuring the frequency. If the resonant frequency is beyond the normal range of the meter, an RF probe must be used.

2. Use some convenient value of capacitance, such as 100, 200, or 300 pF.

3. Adjust the generator output until a convenient mid-scale indication is obtained on the meter. Use an unmodulated signal output from the generator.

4. Starting at a frequency well below the lowest possible resonant frequency of the inductance-capacitance combination under test, slowly increase the generator frequency. If there is no way to judge the approximate resonant frequency, use the lowest generator frequency.

5. Watch the meter for a maximum or peak indication. Note the generator frequency at which the peak indication occurs. This is the resonant frequency of the circuit.

### NOTE

The nature of the peak indication will provide an approximate reflection of the inductor Q or quality. A broad peak indicates a low Q, while a sharp peak indicates a high Q.

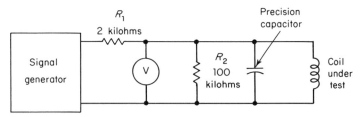

**Fig. 5-24. Inductance measurement with resonant circuit.**

6. Using the resonant frequency and the known capacitance value, calculate the unknown inductance using this equation

$$\frac{\text{inductance}}{(\text{microhenrys})} = \frac{2.53 \times 10^4}{(\text{frequency in megahertz})^2 \times (\text{capacitance in picofarads})}$$

## NOTE

If the amount of distributed capacitance in the coil being tested is small in relation to the known capacitance value, the distributed capacitance can be ignored. Otherwise, the distributed capacitance of the coil under test must be added to the known capacitance value before making the calculation to find inductance.

CHAPTER **6**

# Impedance Measurements

## 6-1. Basic Impedance Equations

The impedance of an a-c circuit is dependent upon the relationship of resistance, inductive reactance, and capacitive reactance. While not all a-c circuits have both inductive and capacitive reactance, all practical a-c circuits do have both resistance and reactance. Therefore, the basic impedance equation for a series circuit is

$$Z(\text{impedance}) = \sqrt{R(\text{resistance})^2 + X(\text{reactance})^2}$$

If there are both $X_C$ (capacitive reactance) and $X_L$ (inductive reactance), and $X_C$ is larger than $X_L$, the equation is

$$Z = \sqrt{R^2 + (X_C - X_L)^2}$$

If there are both $X_C$ and $X_L$, and $X_L$ is larger than $X_C$, the equation is

$$Z = \sqrt{R^2 + (X_L - X_C)^2}$$

At resonance, $X_L = X_C$. Therefore, $Z = R$.

### 6-1.1. Solving Impedances with Voltages and Currents

In practical work, it is often convenient to solve impedance problems using voltage, current, and resistance in their basic Ohm's law relationship.

Reactance $X_C$ or $X_L$ can be substituted in the basic Ohm's law equation for impedance or resistance when calculating the voltage or current of the individual capacitor or inductance but not when calculating the total network (where only impedance, or $Z$, must be used).

For example, in a series LRC circuit (Fig. 6-1), the current through each component is the same. Therefore, if the current is known and the individual reactances are known, the voltage across each component can be calculated using

$$E_R = IR, \qquad E_{X_C} = IX_C, \qquad \text{or} \qquad E_{X_L} = IX_L$$

If the current is not known, a current can be assumed to produce theoretical voltages.

Either way, the theoretical voltages can be added by vector calculation to find the total voltage using

$$E_T = \sqrt{E_R^2 + (E_{X_L} - E_{X_C})^2}$$

or

$$E_T = \sqrt{E_R^2 + (E_{X_C} - E_{X_L})^2}$$

With the total voltage calculated, the total impedance (actual) of the network can be calculated using

$$Z = \frac{E_T}{I}$$

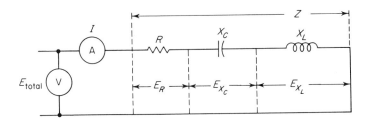

$$E_R = IR, \quad E_{X_C} = IX_C, \quad EX_L = IX_L$$

$$E_{total} = \sqrt{E_R^2 + (E_{X_L} - E_{X_C})^2} \quad \text{or} \quad \sqrt{E_R^2 + (E_{X_C} - E_{X_L})^2}$$

$$\text{also} \quad Z = \frac{E_{total}}{I}, \quad R = \frac{E_R}{I}, \quad X_C = \frac{E_{XC}}{I}, \quad X_L = \frac{E_{XL}}{I}$$

**Fig. 6-1.** Solving series impedances with voltages and currents.

Going further, if the reactance or resistance is not known but the current and voltages are known, the individual reactances or resistance can be calculated using

$$R = \frac{E_R}{I} \quad \text{or} \quad X = \frac{E_X}{I}$$

If neither the voltage nor the reactance is known, the reactance can be calculated using the equations of Fig. 4-5 or Fig. 5-7. This, however, requires that frequency and capacitance or inductance be known.

The steps for solving series LRC impedance problems are shown in Fig. 6-1. Note that these steps can also be applied to a series LR or RC circuit to find impedance.

*In a parallel LRC circuit* (Fig. 6-2), the voltage across each component is the same. Therefore, if the voltage is known, and the individual reactances are known, the current through each component can be calculated using

$$I = \frac{E}{R}, \quad I = \frac{E}{X_C}, \quad \text{or} \quad I = \frac{E}{X_L}$$

If the voltage is not known, a voltage can be assumed to produce theoretical currents.

Either way, the theoretical currents can be added by vector calculation to find the total current using

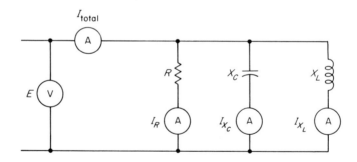

$$I_R = \frac{E}{R}, \quad I_{X_C} = \frac{E}{X_C}, \quad I_{X_L} = \frac{E}{X_L}$$

$$I_{total} = \sqrt{I_R^2 + (I_{X_L} - I_{X_C})^2} \quad \text{or} \quad \sqrt{I_R^2 (I_{X_C} - I_{X_L})^2}$$

$$\text{also} \quad Z = \frac{E}{I_T} \quad R = \frac{E}{I_R} \quad X_C = \frac{E}{I_{X_C}} \quad X_L = \frac{E}{I_{X_L}}$$

**Fig. 6-2.** Solving parallel impedances with voltages and currents.

$$I_T = \sqrt{I_R{}^2 + (I_{X_L} - I_{X_C})^2}$$

or

$$I_T = \sqrt{I_R{}^2 + (I_{X_C} - I_{X_L})^2}$$

With the total current calculated, the total impedance (actual) of the network can be calculated using

$$Z = \frac{E}{I_T}$$

Furthermore, if the reactance or resistance is not known but the voltage and currents are known, the individual resistance or reactances can be calculated using

$$R = \frac{E}{I_R}, \qquad X_L = \frac{E}{I}, \qquad \text{or} \qquad X_C = \frac{E}{I_C}$$

If neither the current nor the reactance is known, the reactance can be calculated using the equations of Fig. 4-5 or Fig. 5-7. However, this requires that frequency and capacitance or inductance be known.

The steps for solving parallel LRC impedance problems are shown in Fig. 6-2. Note that these steps can also be applied to a parallel LR or RC circuit to find impedance.

## 6-2. Effects of Frequency on Impedance

Assuming resistance to be constant, impedance is dependent upon reactance. In turn, reactance is dependent upon frequency. Table 6-1 summarizes the effects of increasing and decreasing frequency on reactance and impedance.

**TABLE 6-1**

Effects of Frequency on Impedance and Reactance

| Circuit or Value | Effect of Frequency | |
| --- | --- | --- |
| | Increase | Decrease |
| Capacitive reactance $(X_C)$ | Lowers $X_C$ | Raises $X_C$ |
| Inductive reactance $(X_L)$ | Raises $X_L$ | Lowers $X_L$ |
| Series resonance $(Z)$ | Raises $Z$ | Raises $Z$ |
| Parallel resonance $(Z)$ | Lowers $Z$ | Lowers $Z$ |
| Series capacitor-resistor combination $(Z)$ | Lowers $Z$ | Raises $Z$ |
| Series inductance-resistor combination $(Z)$ | Raises $Z$ | Lowers $Z$ |
| Parallel capacitor-resistor combination $(Z)$ | Lowers $Z$ | Raises $Z$ |
| Parallel inductance-resistor combination $(Z)$ | Raises $Z$ | Lowers $Z$ |

### 6-3. Matching Impedances

One problem often encountered when testing electronic circuits or components is the matching of impedances. To provide a smooth transition between devices of different characteristic impedance, each device must encounter a total impedance equal to its own characteristic impedance. A certain amount of signal attenuation is usually required to achieve this transition. A simple resistive impedance-matching network that provides minimum attenuation is shown in Fig. 6-3, together with the applicable equations.

For example, to match a 50-ohm system to a 125-ohm system

$$Z_1 = 50 \text{ ohms} \quad \text{and} \quad Z_2 = 125 \text{ ohms}$$

Therefore:

$$R_1 = \sqrt{125\,(125 - 50)} = 96.8 \text{ ohms}$$

$$R_2 = 50\,\sqrt{\frac{125}{125 - 50}} = 64.6 \text{ ohms}$$

Though the network in Fig. 6-3 provides minimum attenuation for a purely resistive impedance-matching device, the attenuation seen from one end does not equal that seen from the other end. A signal applied from the lower impedance source $(Z_1)$ encounters a voltage attenuation $(A_1)$ that may be determined with this equation

$$A_1 = \frac{96.8}{125} + 1 = 1.77 \text{ attenuation}$$

A signal applied from the higher impedance source $(Z_2)$ will produce an even greater voltage attenuation $(A_2)$ that may be determined as fol-

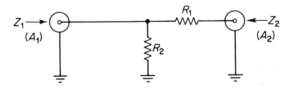

To match impedances: $R_1 = \sqrt{Z_2(Z_2 - Z_1)}$   $R_2 = Z_1\sqrt{\dfrac{Z_2}{Z_2 - Z_1}}$

Voltage attenuation seen from $Z_1$ end $(A_1)$: $A_1 = \dfrac{R_1}{Z_2} + 1$

Voltage attenuation seen from $Z_2$ end $(A_2)$: $A_2 = \dfrac{R_1}{R_2} + \dfrac{R_1}{Z_1} + 1$

**Fig. 6-3.** Resistive impedance-matching network.

lows. Assume that $R_1 = 96.8$ ohms, $R_2 = 64.6$ ohms, and impedance $Z_1 = 50$ ohms.

$$A_2 = \frac{96.8}{64.6} + \frac{96.8}{50} + 1 = 4.44 \text{ attenuation}$$

## 6-4. Measuring Impedance in Power Circuits

### 6-4.1. Current Method

The current measuring circuit of Fig. 2-10 can also be used to measure impedances and power consumption.

If the current and voltage are known, the impedance can be calculated by $Z = E/I$.

The total power consumed (in voltamperes) can be calculated by $VA = E \times I$. Of course, true a-c power (expressed in watts) is found by multiplying the voltamperes by the cosine of the phase angle. The volt-ampere figure can be used for most practical purposes. (Phase angle measurements are discussed in Chapter 8.)

Figure 6-4 shows the basic (current method) test circuit for measuring impedance. Note that the impedance of individual components (head-phones, coils, transformers, etc.) or the impedance presented by a complete circuit or equipment (radio receiver, test equipment, etc.) can be measured using the same basic circuit. The following precautions must be observed when using the current method to find impedance.

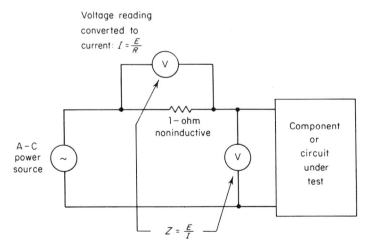

Fig. 6-4. Measuring impedance of component or circuit using current method.

1. Accuracy of measurement is dependent upon accuracy of the resistance value in series with the circuit.

2. A 1-ohm resistor should be used wherever possible. This will simplify the calculations and present the least disturbance to the circuit. If a 1-ohm resistor is not practical (the voltage drop is too small to read), use a 1000-ohm, or 10,000-ohm, resistor. Remember that a 1000-ohm series resistor will convert voltage indications into milliamperes (7 V = 7 mA, etc.).

3. A noninductive series resistance must be used. If the resistor has any inductance (as do most wire-wound resistors) the resistor's inductive reactance will be added to the circuit, producing an error in measurement.

4. The impedance found by the current method applies only to the frequency used during the test. This is no problem for equipment operated from line power (radio receivers, test equipment, etc.) since line power frequency does not change. However, there is a problem for headphones, coils, transformers, etc., which are operated over a wide range of frequencies. Therefore, these components should be tested over the anticipated frequency range. The component or circuit can be checked at various points throughout the frequency range, and any differences in impedance noted.

5. The composition of the waveform will also affect the impedance found using the current method. For example, a nonsine wave or a sine wave containing many harmonics will produce a different impedance reading than will a pure sine wave. From a practical standpoint, it is usually not essential that circuits or components be tested for impedance with a pure sine wave. However, it is essential that the test be made with waveforms equivalent to those used in actual operation.

6. If the impedance and d-c resistance of a component or circuit are known, it is possible to find the reactance, inductance or capacitance, and power factor. These require considerable calculation or vector analysis and are usually of little practical value.

However, the relationship between impedance and d-c resistance can be used to quickly determine the presence of reactance in a supposedly resistive load. For example, a T or L pad used in audio work is considered as a pure resistive load (at audio frequencies). It is possible that such a pad will present some reactive load, especially at the high end of the audio range (15 kHz and above). This condition can be determined by measuring the impedance at the audio frequency, and then comparing its impedance against the d-c resistance. Both values should be substantially the same (allowing for differences due to inaccurate measurements). If there is a large difference between the two values, there is some reactance present. (In the case of an L or T pad, it is most likely inductive reactance.)

### 6-4.2. Resistance-Substitution Method

Figure 6-5 shows the basic-resistance substitution-method test circuit for measuring impedance. Variations of this method to test specific components and circuits are described in later sections and chapters.

The signal source can be an a-c voltage, an RF signal, or a sweep signal as required. The voltage measuring device can be a simple voltmeter, voltmeter with an RF probe, or an oscilloscope. Operation of the circuit is as follows.

1. The signal source is adjusted to the frequency (or frequencies) at which the circuit or component will be operated.

2. The switch $S$ is moved back and forth between position $A$ and position $B$, while resistance $R$ is adjusted until the voltage reading is the same in both positions of the switch.

3. Resistor $R$ is then disconnected from the circuit, and the d-c resistance of $R$ is measured with an ohmmeter. The d-c resistance of $R$ is then equal to the impedance of the circuit or component under test.

The following precautions must be observed when using the resistance-substitution method to find impedance.

1. Accuracy of the impedance measurement is dependent upon the accuracy with which the d-c resistance is measured.

2. A noninductive resistance must be used.

3. The impedance found by the resistance-substitution method applies only to the frequency used during the test.

4. For practical purposes, the test must be made with waveforms equivalent to those used in actual operation.

## 6-5. Measuring Impedance with Pulses

A pulse generator and oscilloscope combination can be used to measure impedance. This impedance measurement method is based on the comparison of reflected pulses with incident or output pulses. As a signal travels down a transmission line, each time the signal encounters a mismatch or different impedance a reflection is generated and sent back along the line to the source. The amplitude and polarity of the reflection are determined by the value of the impedance encountered in relation to the characteristic impedance of the cable.

If the mismatch impedance is higher than that of the line, the reflection will be of the same polarity as the applied signal. If the mismatch impedance is lower than that of the line, the reflection will be of opposite

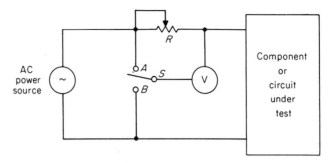

**Fig. 6-5.** Measuring impedance of component or circuit using resistance-substitution method.

polarity. The reflected signal is added to or subtracted from the amplitude of the pulse if the reflected signal returns to the source before the pulse has ended. Thus, for a cable with an open end (no termination), the impedance is infinite, and the pulse amplitude would be doubled. For a cable with a shorted end, the impedance is zero, and the pulse would be canceled.

The basic procedure for measuring impedance with pulses is as follows.

1. Connect the equipment as shown in Fig. 6-6.

2. Place the oscilloscope in operation as described in the instruction manual and switch on the internal recurrent sweep. Set the sweep selector and sync selector to internal.

3. Switch on the pulse generator as described in the instruction manual. Set the sweep frequency and sync controls to display the output pulse and the *first* reflected pulse, as shown in Fig. 6-6.

4. Observe the output and reflected pulses on the oscilloscope screen. Using Fig. 6-6 as a guide, determine the values of $V_o$ (output voltage amplitude) and $V_x$ (reflected voltage amplitude).

5. Calculate the unknown impedance using the following equation

$$Z = \frac{50}{\left(\dfrac{2\,V_o}{V_x}\right) - 1}$$

where  $Z$ is unknown impedance,
       50 is reference impedance (50-ohm coaxial line),
       $V_o$ is peak amplitude produced by the 50-ohm reference impedance, and
       $V_x$ is peak amplitude at the time of first reflection.

6. In the example of Fig. 6-6, the reflected pulse voltage $V_x$ is lower in amplitude than the output pulse voltage $V_o$. This indicates that the re-

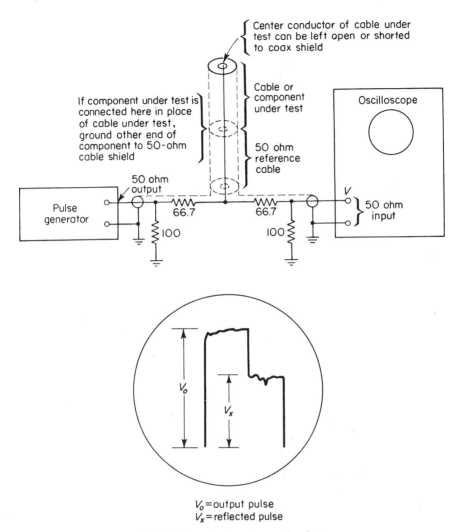

$V_o$ = output pulse
$V_x$ = reflected pulse

**Fig. 6-6.** Measuring impedance with pulses.

flected pulse is of opposite polarity (subtracting from the output pulse) and that the impedance of the component or cable under test is lower than that of the reference cable. For example, assume that $V_o$ is 20 V and $V_x$ is 16 V (with a 50-ohm reference cable). Using this equation

$$Z = \frac{50}{\left(\frac{2 \times 20}{16}\right) - 1} = 33.3 \text{ ohms impedance}$$

## NOTE

The test connection diagram of Fig. 6-6 can be used to measure the impedance of either a cable or component. If a cable is being tested and the cable is long in relation to the reference cable, the open end of the cable can be terminated or left open, whichever is convenient. There will be a second reflection from the cable open end that may be confused with the first reflection. However, a long cable will produce the second reflection far removed from the first reflection. If a component is being tested, and the component has an open end, this end should be grounded as shown in Fig. 6-6.

## 6-6. Measuring Impedances with Sweep Frequencies

One of the main problems in measuring impedance with the procedures described in other sections of this chapter is that the impedance values are valid for *one frequency only*. The tests must be repeated over a wide range of frequencies (unless the device being tested is to be used at one frequency only). This problem can be minimized (if not eliminated) by means of sweep frequency techniques using a sweep generator and an oscilloscope. The sweep frequency technique permits the impedance to be measured over a wide band of frequencies simultaneously.

A sweep generator is an FM generator (a frequency-modulated radio-frequency generator). When a sweep generator is set to a given frequency, this is the *center frequency*. The sweep generator is tuned to sweep the band of frequencies at which the impedance tests are to be made. A trace representing the response characteristics of the device under test is displayed on the oscilloscope. The trace is synchronized with the generator sweep rate so that any point on the oscilloscope trace corresponds to the instantaneous frequency of the sweep generator.

Some sweep generators incorporate a marker generator; others require an external marker generator. Basically, a marker generator is an RF signal generator that has highly accurate dial markings and can be calibrated precisely. The marker generator is used to provide calibrated markers along the oscilloscope trace (or response curve of the device under test). When the marker signal from the marker generator is coupled into the test circuit, a vertical "pip" or marker appears on the trace. When the marker generator is tuned to a frequency within the passband accepted by the device under test, the marker indicates the position of that frequency on the sweep trace.

Some sweep generators incorporate a blanking circuit. When the sweep generator output is swept across its band of frequencies, the frequencies go from low to high, then return from high to low. With the blanking circuit actuated, the return or retrace is blanked off. This makes it possible to view a zero-reference line on the oscilloscope during the retrace period.

### 6-6.1. Basic Sweep Frequency Technique

The following steps describe the *basic sweep frequency technique.* Later paragraphs of this section show how the basic technique can be adapted to impedance measurements. Later chapters describe how the sweep frequency technique can be used for measurement of frequency and test of components.

1. Connect the equipment as shown in Fig. 6-7.
2. Place the oscilloscope and sweep generator in operation as described in the related instruction manuals.

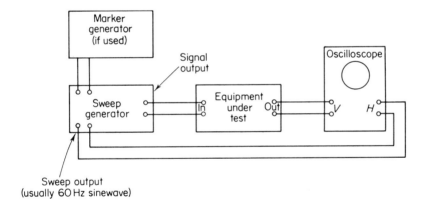

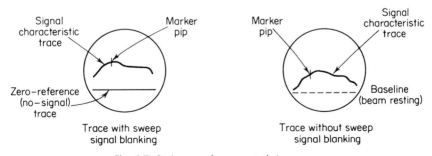

**Fig. 6-7. Basic sweep frequency technique.**

3. Switch off the oscilloscope internal recurrent sweep. Set the oscilloscope sweep selector and sync selector to external.

4. Under these conditions, the oscilloscope horizontal sweep can be obtained from the generator sweep output. The length of the horizontal sweep should represent the total sweep spectrum. For example, if the sweep is from 10 to 20 kHz, the left-hand end of the horizontal trace will represent 10 kHz, and the right-hand end will represent 20 kHz. Any point along the horizontal trace will represent a corresponding frequency. For example, the midpoint on the trace would represent 15 kHz. If a rough approximation of frequency is desired, the oscilloscope horizontal gain control can be adjusted until the trace occupies an exact number of scale divisions, such as 10 cm for the 10- to 20-kHz sweep signal. Each centimeter division would then represent 1 kHz.

5. If a more accurate frequency measurement is desired, the marker generator must be used. The marker generator output frequency is adjusted until the marker pip is aligned at the desired point on the trace. The frequency is then read from the marker generator frequency dial.

6. The response curve (trace) depends upon the device under test. If the device has a passband and the sweep generator is set so that its sweep is wider than the passband, the trace will start low (in amplitude and vertical deflection), rise toward the middle, and then drop off at the right, as shown in Fig. 6-7. The sweep frequency technique will tell at a glance the overall passband characteristics of the device (sharp response, flat response, irregular response at certain frequencies, etc.). The exact frequency limits of the passband can be measured with the marker-generator pip.

7. Switch the sweep generator blanking control on or off as desired. (Some sweep generators do not have a blanking function.) With the blanking function in effect, there will be a zero-reference line on the trace. With the blanking function off, the horizontal baseline will not appear. The sweep generator blanking function is not to be confused with the oscilloscope blanking (which is bypassed when the sweep signal is applied to the horizontal amplifier).

## NOTE

The basic sweep frequency technique can be performed with the oscilloscope sweep selector set to internal and the sync selector set to line. Two conditions must be met. First, the sweep generator must be swept at the line frequency (usually 60 Hz). Second, the oscilloscope or sweep generator must have a *phasing control* so that the two sweeps can be synchronized. If there is a phase shift between the sweep generator and the oscilloscope horizontal sweep, even though they are at the same frequency, a double pattern will appear. This condition can be corrected by shifting the oscilloscope sweep drive signal phase with the phasing control. The alternate method is used when the sweep generator does not have a sweep output separate from the signal output or when it is not desired to use the sweep output. Blanking of the trace (if any blanking is used) is controlled by the oscilloscope circuits.

### 6-6.2. Checking Sweep Generator Output Uniformity

Many sweep generators do not have a uniform output. That is, the output voltage is not constant over the swept band. This can lead to false conclusions in some tests. In other tests, it is only necessary to know the amount of nonuniformity and make allowances. For example, a sweep generator can be checked for flatness before connection to a circuit, and any variation in output noted. If the output remains the same after it is connected to the circuit, even though the output shows variations, the circuit under test is not at fault. Sweep generator output can be checked as follows.

1. Connect the equipment as shown in Fig. 6-8.

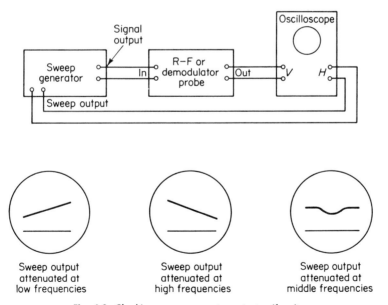

Fig. 6-8. Checking sweep generator output uniformity.

## NOTE

The probe shown in Fig. 6-8 can be omitted if the sweep generator output signal frequency is within the passband of the oscilloscope vertical amplifier. This is not usually the case with shop-type oscilloscopes, therefore some form of demodulator is necessary. If the oscilloscope is not equipped with a demodulator or RF probe, demodulator networks can be fabricated as shown in Fig. 6-9. Figure 6-9a is for a single-ended sweep generator, such as the type usually found in laboratory work. Figure 6-9b is for a double-ended or balanced output. The balanced output is often found on sweep generators specifically for use with TV and FM receivers.

2. Place the oscilloscope and sweep generator in operation as described in the related instruction manual.

3. Switch off the oscilloscope internal recurrent sweep. Set the oscilloscope sweep selector and sync selector to external so that the horizontal sweep is obtained from the generator sweep output.

4. Set the sweep generator width to maximum. Switch the sweep generator blanking control on or off as desired.

5. With the blanking function not in effect, only one trace will appear. This should be deflected vertically from the normal trace resting position.

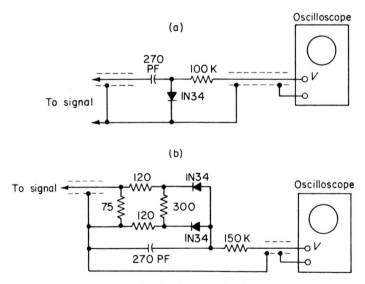

Fig. 6-9. RF demodulator probe circuits.

With the blanking function on, there will be two traces. The upper trace is the generator output characteristic; the lower trace is the no-signal trace.

6. Check the sweep generator output trace for flatness. A drop or slope of the right-hand side of the trace indicates that the sweep output is reduced at the high-frequency end of the sweep. A slope to the left indicates a reduced output at the low-frequency end of the sweep. A sudden dropping off of the trace or a dip in the middle indicates an uneven output. A perfectly flat trace or, more realistically, a trace that has only slight curvature at the ends, indicates an even output across the entire swept band.

7. Leave the sweep width at maximum, but adjust the center frequency of the sweep generator over its entire range. Check that the output is flat, or at least that any variations are consistent, across the range of the sweep generator.

### 6-6.3. Measuring Transmission Line Impedance

The sweep frequency technique can be combined with the resistance-substitution method (Section 6-4.2) to measure the impedance of transmission lines over a broad frequency range. The same technique can be applied to TV antenna lead-in wire, coaxial cables, or any line that is supposed to have a constant impedance along its entire length. (Transmission line impedance can also be measured using other methods such as the dip circuit technique described in Chapter 11.) The procedure for sweep frequency measurement of a transmission line is as follows.

1. Connect the equipment as shown in Fig. 6-10.

2. Place the oscilloscope in operation as described in the instruction manual. Set the oscilloscope sweep selector and sync selector to external.

3. Place the sweep generator in operation as described in the instruction manual. Switch sweep generator blanking control on or off as desired. Tune the sweep generator to the normal operating frequency at which the transmission line is to be used. Adjust the sweep width to cover the complete range of frequencies.

4. Adjust the variable resistance to the supposed impedance value of the transmission line.

5. The oscilloscope pattern should show a flat trace. If it does not, adjust the variable resistance until the trace is flat.

6. Disconnect the variable resistance (without disturbing its setting) and measure the resistance with an ohmmeter. This value is equal to the transmission line characteristic impedance.

7. If the trace cannot be made flat, temporarily disconnect the transmission line but leave the sweep generator output connected to the oscillo-

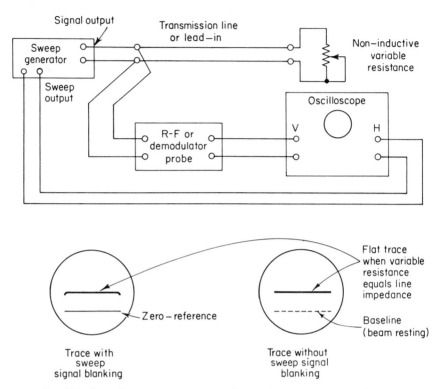

**Fig. 6-10.** Measuring transmission line impedance with sweep frequency technique.

scope. Check the trace pattern. If it is still not flat, the sweep generator output is not uniform. (Refer to Section 6-6.2.) If the trace is flat with the line disconnected but cannot be made flat with any setting of the variable resistance, the transmission line impedance is not uniform.

### 6-6.4. Measuring Impedance Match

The sweep frequency technique can be combined with a form of reflected measurement similar to that described in Section 6-5 to measure impedance match. This method is particularly effective when the main concern is with match or mismatch, not with actual impedance value. For example, the method will provide a quick check of impedance match between a transmission line and an antenna (or any other terminating device) over a broad frequency range. The procedure for sweep frequency measurement of impedance match is as follows.

1. Connect the equipment as shown in Fig. 6-11.

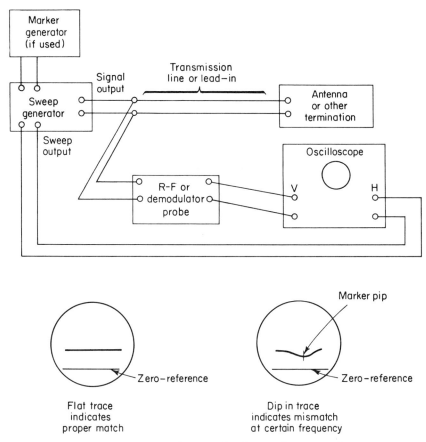

Fig. 6-11. Measuring impedance match with sweep frequency technique.

2. Place the oscilloscope in operation as described in the instruction manual. Set the oscilloscope sweep selector and sync selector to external.

3. Place the sweep generator in operation as described in the instruction manual. Switch the sweep generator blanking control on or off as desired. Tune the sweep generator to the normal operating frequency with which the transmission line and antenna (or other terminating device) are to be used. Adjust the sweep width to cover the complete range of frequencies.

4. If the transmission line and antenna are properly matched, the oscilloscope will show a flat trace. If the oscilloscope trace is not flat and it is desired to find the frequency at which the mismatch occurs, the marker generator can be adjusted until the marker pip is aligned at the desired

point on the trace. The mismatch frequency, or band of frequencies, can be read from the marker generator frequency dial.

5. If there is any doubt that the variation in trace flatness is caused by variation in sweep generator output, temporarily disconnect the transmission line but leave the sweep generator output connected to the oscilloscope. If the variation is removed or reduced drastically when the transmission line and antenna (or other termination) are disconnected, a mismatch is indicated.

### 6-6.5. Measuring Mismatch and Attenuation of Devices Inserted in Transmission Lines

When any device is inserted in a transmission line, there occurs the possibility of a mismatch in impedance between the line and the device. This will produce attenuation at certain frequencies. Even if there is no mismatch, the device (such as a coupler, splitter, pick-off probe) can produce some attenuation at all frequencies. The sweep frequency technique can be combined with a form of reflected measurement to check impedance match and attenuation of any device inserted into a transmission line, as follows.

1. Connect the equipment as shown in Fig. 6-12.

2. Place the oscilloscope in operation as described in the instruction manual. Set the oscilloscope sweep selector and sync selector to external.

3. Place the sweep generator in operation as described in the instruction manual. Switch the sweep generator blanking control on or off as desired. Tune the sweep generator to the normal operating frequency with which the transmission line and inserted device are to be used. Adjust the sweep width to cover the complete range of frequencies.

4. With the probe connected at the sweep generator output, the oscilloscope will show a flat trace if the transmission line and inserted device are properly matched. If the trace is not flat and it is desired to find the frequency at which the mismatch occurs, the marker generator can be adjusted until the marker pip is aligned at the desired point on the trace. The mismatch frequency (or band of frequencies) can be read from the marker generator frequency dial.

5. If there is any doubt that the variation in trace flatness is caused by variation in sweep generator output, temporarily disconnect the transmission line but leave the oscilloscope probe connected to the sweep generator output. If the variation is removed or drastically reduced when the transmission line and inserted device are disconnected, a mismatch is indicated.

6. Once the match or mismatch has been established, connect the

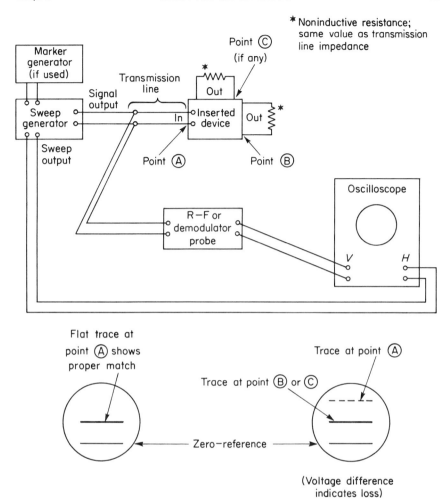

**Fig. 6-12.** Measuring mismatch and attenuation of devices inserted in transmission lines with sweep frequency technique.

probe to the inserted device input at point $A$ (Fig. 6-12). This will establish a vertical reference deflection. Then, connect the probe to point $B$ and to point $C$ (if any) in turn.

7. The oscilloscope vertical deflection should be lower at points $B$ and $C$ than at point $A$ because almost any device inserted into a transmission line will show some loss or attenuation. Unless there is some special design characteristic, the output at points $B$ and $C$ should be the same. The voltage difference between the signal at points $A$ and $B$ (or $A$ and $C$) can be measured directly on the oscilloscope (assuming that the oscillo-

scope vertical system is voltage calibrated) and converted to a ratio, decibel value, or whatever is desired.

### 6-6.6. Measuring Input and Output Impedances

The sweep frequency technique can be combined with the resistance-substitution method (Section 6-4.2) to measure the input and output impedances of RF (and audio) components over a broad range of frequencies. The procedure is as follows.

1. Connect the equipment as shown in Fig. 6-13.
2. Place the oscilloscope in operation as described in the instruction manual. Set the oscilloscope sweep selector and sync selector to external.
3. Place the sweep generator in operation as described in the instruction manual. Switch the sweep generator blanking control on or off as desired. Tune the sweep generator to the normal operating frequency at which the component is to be used. Adjust the sweep width to cover the complete range of frequencies.

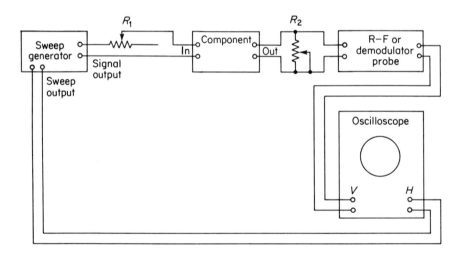

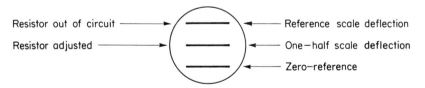

**Fig. 6-13.** Measuring input and output impedances with sweep frequency technique.

4. Disconnect resistors $R_1$ and $R_2$ from the circuit.

5. Adjust the sweep generator output level so that the trace is at a convenient vertical scale marking. Note this vertical deflection scale marking.

6. Connect resistor $R_1$ back into the circuit. Vary the resistance of $R_1$ until the voltage indicated on the oscilloscope is one-half the original value.

7. Disconnect resistor $R_1$ from the circuit and measure its d-c resistance. This resistance is equivalent to the input impedance of the component.

8. With both $R_1$ and $R_2$ out of the circuit, again adjust the sweep generator output level so that the trace is at a convenient vertical scale indication.

9. Connect resistor $R_2$ back into the circuit. Vary the resistance of $R_2$ until the voltage indication on the oscilloscope is one-half the original value.

10. Disconnect resistor $R_2$ from the circuit and measure its d-c resistance. This resistance is equivalent to the output impedance of the component.

## NOTE

If the input or output impedance contains some reactive component, the impedance will change as the frequency is swept across the band. This may make it difficult to adjust the resistors $R_1$ and $R_2$ to produce a half-scale value. (The trace may slope or vary.) In that case, pick out some point on the reference trace (resistor out of the circuit) and then adjust the resistor so that the same point is set to one-half scale.

# CHAPTER 7

# Decibel Measurements

## 7-1. Power Ratio Units

The decibel, or dB, is the unit that has been widely adopted in electronics to express logarithmically the ratio between two power or voltage levels and less commonly the ratio between two current levels. A decibel is one-tenth of a bel. (The bel is too large for most practical applications.)

Although power, voltage, or current amplification or the magnitude of a particular power, voltage, or current relative to a given reference value can be expressed as an ordinary ratio, the decibel has been adopted because of its much greater convenience.

Because the response of the human ear to sound waves is *approximately* proportional to the logarithm of the energy of the sound wave and is not proportional to the energy itself, the use of a logarithmic unit permits a closer approach to the reaction of the human ear. The impression gained by the human ear as to the magnitude of sound is roughly proportional to the logarithm of the actual energy contained in the sound. For example, the change in gain of an amplifier expressed in decibels provides a much better index of the effect of the sound upon the ear than it does if expressed as a power or voltage ratio.

Mathematically the decibel is a function of the following.

$$10 \log \frac{P_2}{P_1}$$

Usually, $P_2$ represents power output and $P_1$ represents power input. Therefore, if $P_2$ is greater than $P_1$, there is a power gain expressed in decibels. With $P_1$ greater than $P_2$, there is a power loss expressed in negative decibels. Whichever is the case, the ratio of the two powers ($P_1$ and $P_2$) is taken, and the logarithm of this ratio is multiplied by 10. The mathematical expression need not be used if it is remembered that *doubling of a power represents 3 dB*.

The decibel is also the function of

$$20 \log \frac{E_2}{E_1} \qquad 20 \log \frac{I_2}{I_1}$$

Thus, the ratio of two voltages (or currents) is taken, and the logarithm of this ratio is multiplied by 20.

Although power ratios are independent of source and load impedance values, voltage and current ratios in these equations hold true only when the source and load impedances are equal. In circuits where these impedances differ, voltage and current ratios are calculated as follows:

$$20 \log \frac{E_1 \sqrt{R_2}}{E_2 \sqrt{R_1}} \qquad 20 \log \frac{I_1 \sqrt{R_1}}{I_2 \sqrt{R_2}}$$

where $R_1$ is source impedance and
　　　$R_2$ is load impedance.

($E_1 \sqrt{R_2}$ and $I_1 \sqrt{R_1}$ are always higher in value than $E_2 \sqrt{R_1}$ and $I_2 \sqrt{R_2}$.)

### 7-1.1. Volume Units and dBM

Decibels are often used with specific reference levels. The most common reference levels in use are the *volume unit,* or VU, and the *decibel meter,* or dBM.

When the VU is used, it is assumed that the zero level is equal to 0.001 watt (1 mW) across a 600-ohm impedance. Therefore

$$VU = 10 \log \frac{P_2}{0.001} = 10 \log \frac{P_2}{10^{-3}} = 10 \log 10^3 P_2$$

Since the log of $10^3$, or 1000, is 3, the equation can be simplified to

$$VU = 30 \log P_2$$

Both the dBM and VU have the same zero level base. A dBM scale is (generally) used when the signal is a sine wave (normally 1 kHz), whereas the VU is used for complex audio waveforms.

### 7-1.2. dBW, dBV, dBA, and dBRN

In addition to dBM and VU, there are four other decibel units. Their use is normally limited to audio or sound work.

1. dBW. Whereas the dBM unit has a zero level of 1 mW, the dBW has a zero level of 1 W. When the dBW unit is used, decibels above or below this level are termed $\pm$dBW.

$$1 \text{ dBW} = 30 \text{ dBM}$$

2. dBV. The dBV expresses the response of a microphone at a given frequency in dB. A reference level of 0 dB is equal to 1 V (dBV) when dyn/cm$^2$ of sound pressure is exerted on a microphone. When the dBV is used, decibels above or below the reference level are termed $\pm$dBV.

3. dBA. The dBA is used to express the relationship between the noise interference produced by a noise frequency (or a band of noise frequencies) and a standard reference-noise power level. Actually, two standard reference levels are used. One has been established as $-90$ dBM ($10^{-12}$ W) at a frequency of 1 kHz using a Western Electric Type 144 handset as the standard equipment. Later another reference level was established as $-85$ dBM ($10^{-11.5}$ W), using an improved handset, the Western Electric Type F1A. Therefore, the dBA has two standard reference (zero) levels, $-90$ dBM and $-85$ dBM. When this unit is used, the number of the handset employed is given to identify the reference level.

4. dBRN. The dBRN is numerically identical to the dBA, having a reference level of $-90$ dBM, as established using the Type 144 handset.

### 7-1.3. Neper

The neper (Np), like the decibel, is used to measure differences in power level. However, the decibel is based on common logarithms to the base 10, whereas the neper is based on Napierian logarithms to the base $\epsilon$ (2.718281). The neper is not generally used in English-speaking countries. When used, the neper is

$$\frac{1}{2} \log_\epsilon \frac{P_2}{P_1}$$

The following relationship exists between the neper and dB.

$$1 \text{ dB} = 0.1151 \text{ Np} \qquad 1 \text{ Np} = 8.686 \text{ dB}$$

## 7-2. Decibel Conversion Chart

Table 7-1 lists the conversion factor for a number of power ratios and voltage or current ratios. These ratios are carried out to four places for maximum accuracy and ease of interpolation.

**TABLE 7-1**

Decibel Conversion Chart

| Power Ratio | Voltage and Current Ratio | Decibels (+) (−) | Voltage and Current Ratio | Power Ratio |
|---|---|---|---|---|
| 1.000 | 1.000 | 0.0 | 1.000 | 1.000 |
| 1.023 | 1.012 | 0.1 | .9886 | .9772 |
| 1.047 | 1.023 | 0.2 | .9772 | .9550 |
| 1.072 | 1.035 | 0.3 | .9661 | .9333 |
| 1.096 | 1.047 | 0.4 | .9550 | .9120 |
| 1.122 | 1.059 | 0.5 | .9441 | .8913 |
| 1.148 | 1.072 | 0.6 | .9333 | .8710 |
| 1.175 | 1.084 | 0.7 | .9226 | .8511 |
| 1.202 | 1.096 | 0.8 | .9120 | .8318 |
| 1.230 | 1.109 | 0.9 | .9016 | .8128 |
| 1.259 | 1.122 | 1.0 | .8913 | .7943 |
| 1.585 | 1.259 | 2.0 | .7943 | .6310 |
| 1.995 | 1.413 | 3.0 | .7079 | .5012 |
| 2.512 | 1.585 | 4.0 | .6310 | .3981 |
| 3.162 | 1.778 | 5.0 | .5623 | .3162 |
| 3.981 | 1.995 | 6.0 | .5012 | .2512 |
| 5.012 | 2.239 | 7.0 | .4467 | .1995 |
| 6.310 | 2.512 | 8.0 | .3981 | .1585 |
| 7.943 | 2.818 | 9.0 | .3548 | .1259 |
| 10.00 | 3.162 | 10.0 | .3162 | .10000 |
| 12.59 | 3.548 | 11.0 | .2818 | .07943 |
| 15.85 | 3.981 | 12.0 | .2515 | .06310 |
| 19.95 | 4.467 | 13.0 | .2293 | .05012 |
| 25.12 | 5.012 | 14.0 | .1995 | .03981 |
| 31.62 | 5.632 | 15.0 | .1778 | .03162 |
| 39.81 | 6.310 | 16.0 | .1585 | .02512 |
| 50.12 | 7.079 | 17.0 | .1413 | .01995 |
| 63.10 | 7.943 | 18.0 | .1259 | .01585 |
| 79.43 | 8.913 | 19.0 | .1122 | .01259 |
| 100.00 | 10.000 | 20.0 | .1000 | .01000 |
| $10^3$ | 31.62 | 30.0 | .03162 | .00100 |
| $10^4$ | $10^2$ | 40.0 | $10^{-2}$ | $10^{-4}$ |
| $10^5$ | 316.23 | 50.0 | $3.162 \times 10^{-3}$ | $10^{-5}$ |
| $10^6$ | $10^3$ | 60.0 | $10^{-3}$ | $10^{-6}$ |
| $10^7$ | $3.162 \times 10^3$ | 70.0 | $3.162 \times 10^{-4}$ | $10^{-7}$ |
| $10^8$ | $10^4$ | 80.0 | $10^{-4}$ | $10^{-8}$ |
| $10^9$ | $3.162 \times 10^4$ | 90.0 | $3.162 \times 10^{-5}$ | $10^{-9}$ |
| $10^{10}$ | $10^5$ | 100.0 | $10^{-5}$ | $10^{-10}$ |

## 7-3. Practical Decibel Measurements

Most VOMs (and a few electronic voltmeters) are provided with decibel scales. Decibel measurements are usually made with a *blocking capacitor* (part of the meter circuit) in series with one of the test leads.

The test lead on a typical VOM is connected to a terminal marked "output" or some similar function. This blocks any dc in the circuit being measured from passing to the meter circuit. Such dc may or may not damage the meter, depending upon conditions.

Once the connection has been made, the a-c voltage circuit of the meter is used in the normal manner, except that the readout is made on the decibel scales. Inexperienced operators are often confused by the decibel scales, and the following notes should clarify their use.

The decibel scales represent *power ratios,* not voltage ratios. In most cases, 0 dB is considered as 1 mW (0.001 W) of power across a 600-ohm pure resistive load. This also represents 0.775 V RMS across a 600-ohm pure resistive load.

The decibel scale is related directly to one of the a-c scales, usually the lowest scale on the meter. The VOM range selector must be set to that a-c scale if readings are to be *taken directly* from the decibel scale. If another a-c scale is selected by the range selector, a certain decibel value must be added to the indicated decibel scale value.

For example, in a typical VOM the decibel scale is related directly to the 3-V a-c scale. If the range selector is set to 3 V ac, the decibel scale may be read out directly. (Note that 0 dB is aligned with the 0.775-V point on the directly related a-c scale.)

If the range selector were set to 8, 40, or 160 V ac (the ranges of a typical VOM), it would be necessary to add 3.5, 22.5, or 34.5 dB to the indicated decibel scale reading. The decibel correction values are printed on the meter face and *are applicable to that meter only.* Always consult the meter face (or instruction manual) for data regarding the decibel scales.

## NOTE

The decibel scale readings of any meter *will not* be accurate if the voltages are other than pure sine waves, the load impedances are other than pure resistive, and the load is other than 600 ohms.

### 7-3.1. Correction Factors for Decibel Measurements

If the load is other than 600 ohms, it is possible to apply a correction factor. The decibel is based on the mathematical function of

$$1 \text{ dB} = 10 \log \frac{P_2}{P_1}$$

Since the power will change by a corresponding ratio when resistance is changed (power will increase if resistance decreases and voltage remains the same), it is possible to convert the factor on the right-hand side of the equation to 10 log $R_2/R_1$, where $R_2$ is 600 ohms and $R_1$ is the resistance value of the actual load.

For example, assume that the load resistance is 500 ohms instead of 600 ohms, and a 0-dB indication is obtained (0.775 V RMS).

Divide 600 by 500 = 1.2
The log of 1.2 is 0.0792
10 times 0.0792 = 0.792

Therefore, 0.792 (or 0.8 for practical purposes) must be added to the 0-dB value to give a true value of 0.8 dB.

Table 7-2 lists correction factors to be applied for some common load impedance values. Table 7-2 shows the amount of decibel correction to be added to the indicated decibel value, when the load impedance is other than 600 ohms. For example, if the load impedance is 300 ohms, +3 dB must be added to the indicated value. This can be verified using the following method.

Divide 600 by 300 = 2
The log of 2 is 0.3010
10 times 0.3010 = 3.01 (or 3.0 for practical purposes).

Figure 7-1 shows the relationship between a-c voltages (RMS), decibels, and power (in milliwatts across a 600-ohm pure resistive load). This illustration can be used when a particular meter does not have a decibel scale but does show RMS voltages. Of course, the correction factor of Table 7-2 must be applied to the value if the load is not 600 ohms.

For example, assume that a 2.5-V RMS signal were measured across a 50-ohm load. Figure 7-1 shows 2.5-V RMS equals +10 dB (actually slightly higher).

**TABLE 7-2**

Correction Factors for dB Readings Across Loads
Other Than 600 Ohms

| Resistive Load at 1 kHz | dBM |
|---|---|
| 500 | +0.8 |
| 300 | +3.0 |
| 250 | +3.8 |
| 150 | +6.0 |
| 50 | +10.8 |
| 15 | +16.0 |
| 8 | +18.8 |
| 3.2 | +22.7 |

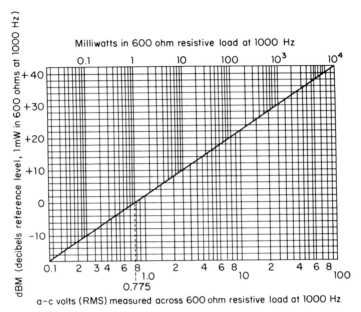

**Fig. 7-1.** Graph for conversion of RMS voltages to dBM values. (Courtesy RCA)

Table 7-2 shows that a +10.8-dB correction factor must be added, resulting in a 20.8-dB, or possibly a 21-dB, true reading.

Note that if the load resistance is greater than 600 ohms, the power will be reduced, and the correction factor must be *subtracted*.

Note also that dBM values are usually based on a frequency of 1 kHz, since this is the frequency at which the decibel system most nearly corresponds to the characteristics of the human ear.

### 7-3.2. Decibel Gain Measurements

There is often confusion in making decibel measurements at the input and output of a particular circuit (such as an amplifier) to find gain or loss, particularly in regard to load impedances. The following rules should clarify this problem.

*If the input- and output-load impedances are 600 ohms* (or whatever value is used on the meter scale) there should be no problem. Simply make a decibel reading at the input and at the output (under identical conditions), subtract the smaller decibel reading from the larger, and note the decibel gain (or loss).

For example, assume that the input shows 3 dB, with 13 dB at the output. This represents a 10-dB *power* gain. If the output has been 3 dB, with 13-dB input, there would be a 10-dB *power* loss.

The decibel gain (or loss) can be converted into a power ratio (or voltage or current ratio) by means of Table 7-1. In the example of a 10-dB gain (or loss) this represents a power ratio of 10 and a voltage or current ratio of 3.1623.

*If the input and output load impedances are not 600 ohms but are equal to each other,* then the *relative* decibel gain or loss is correct, even though the absolute decibel reading is incorrect.

For example, assume that the input and output load impedances are 50 ohms and the input shows 3 dB, with 13 dB at the output. Table 7-2 shows that 10.8 dB must be added to the input and output readings to obtain the correct dBM *absolute* value. However, there is still a 10-dB difference between the two readings. Therefore, the circuit shows a 10-dB gain, and the power (or voltage/current) ratios of Table 7-1 still hold true.

*If the input and output load impedances are not equal,* the relative dB gain or loss indicated by the meter scales will be incorrect.

For example, assume that the input impedance is 300 ohms, the output impedance is 8 ohms, and the input shows $+7$ dB while the output shows $+3$ dB on the scales of a meter using a 600-ohm reference.

There is an apparent loss of 4 dB (input 7 dB — output 3 dB). However, by reference to Table 7-2, it will be seen that the 300-ohm input (7 dB) requires a correction of $+3$ dB, giving a corrected input of $+10$ dB, and the 8-ohm output (3 dB) requires a correction of $+18.8$ dB, giving a correct output of $+21.8$ dB. Thus, there is an actual gain of $+11.8$ dB.

## 7-4. Effect of Blocking Capacitor on Decibel Readings

When the decibel scales are selected on most meters, a blocking capacitor is connected in series with the test lead to prevent dc from entering the meter circuit. In some cases, the blocking capacitor and meter resistance form a high-pass filter and may attenuate low-frequency a-c voltages. However, most meters will provide accurate decibel indications above 15 or 20 Hz.

It is also possible that the blocking capacitor and meter movement coil will form a resonant circuit and increase the decibel readings at some particular frequency (usually about 30 to 60 kHz). Therefore, always consider the frequency problem when making any decibel measurements.

### 7-4.1. Checking Effect of Frequency on Decibel Measurements

The effect of the blocking capacitor on decibel readings at various frequencies can be checked using the following procedure. With the effect (if any) recorded, a *scale factor* can be applied to the readings as necessary.

1. Connect the meter to an audio generator as shown in Fig. 7-2.

2. Set the meter to measure a-c voltage directly, not through the blocking capacitor.

3. Set the audio generator to some low frequency near the low end of the meter's rated frequency range (20 Hz, 60 Hz, etc.).

4. Adjust the amplitude of the audio generator signal so that the meter indicates some a-c voltage near the mid-scale.

5. Without changing any of the audio generator or meter controls, apply the generator output through the "output" or decibel measurement function (with blocking capacitor).

6. If there is no change in voltage indication, the blocking capacitor has no effect on the circuit. However, if the reading is lower with the blocking capacitor in the circuit (as is usually the case), a scale factor must be applied.

7. For example, assume that the reading is 7 V on normal ac (without blocking capacitor) and 5 V with the blocking capacitor. Then a $\frac{7}{5}$ scale factor must be applied to all readings made with the blocking capacitor. (Divide the readings obtained with the blocking capacitor by five and then multiply by seven.)

### NOTE

If there is any d-c voltage at the audio generator output terminals, use an isolating capacitor as shown by dotted line in Fig. 7-2.

8. Reconnect the meter to measure a-c voltage directly, not through the blocking capacitor.

9. While maintaining the audio generator amplitude constant, vary the frequency from the low end up to about 100 kHz. Note and record the voltage reading at various frequencies (say, every 5 kHz).

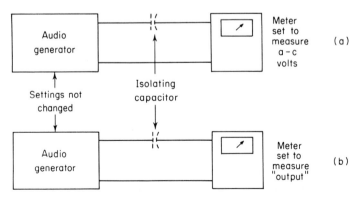

**Fig. 7-2.** Checking effect of "output" function on a-c voltage readings.

10. Repeat Step 9 with the audio generator output applied through the blocking capacitor. Compare the voltage readings at each of the frequencies. The readings should be substantially the same from about 100 Hz and above (except for a possible increase in reading when the blocking capacitor and meter movement form a resonant circuit, as previously discussed). If there are any substantial differences in voltage readings, make a note of the frequencies at which the differences occur. Such a record can be especially useful with decibel gain measurements of audio amplifiers.

# CHAPTER 8

# Time, Frequency, and
# Phase Measurement

## 8-1. Relationship of Wavelength, Frequency, Period, and Velocity

Alternating current changes its polarity at specific intervals as shown in Fig. 8-1. The distance from the point where the voltage or current crosses zero and starts in the positive direction, to the point where the voltage or current just returns to zero from the negative direction, is considered one *wavelength*. A wavelength could also be considered the distance between successive positive peaks, successive negative peaks, or any corresponding points along the waveform. Wavelength is usually expressed in meters (or submultiples of a meter).

The number of wavelengths per unit of time (usually 1 second) is the *frequency* of the alternating current. Frequency is expressed in *hertz* (which has replaced the term *cycles*). Time units are implied with the term hertz. For example, megahertz means megacycles per second.

The time duration of a single wavelength is the *period* of the alternating current. The period is usually expressed in seconds.

The calculations for wavelength, frequency, and period are given in Fig. 8-1. These equations apply to alternating current at all frequencies (AF, RF, etc.).

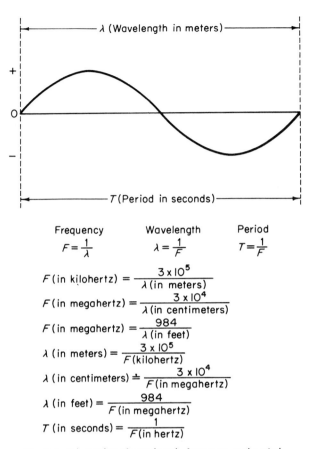

Frequency $\qquad$ Wavelength $\qquad$ Period

$$F = \frac{1}{\lambda} \qquad \lambda = \frac{1}{F} \qquad T = \frac{1}{F}$$

$$F\,(\text{in kilohertz}) = \frac{3 \times 10^5}{\lambda\,(\text{in meters})}$$

$$F\,(\text{in megahertz}) = \frac{3 \times 10^4}{\lambda\,(\text{in centimeters})}$$

$$F\,(\text{in megahertz}) = \frac{984}{\lambda\,(\text{in feet})}$$

$$\lambda\,(\text{in meters}) = \frac{3 \times 10^5}{F\,(\text{kilohertz})}$$

$$\lambda\,(\text{in centimeters}) = \frac{3 \times 10^4}{F\,(\text{in megahertz})}$$

$$\lambda\,(\text{in feet}) = \frac{984}{F\,(\text{in megahertz})}$$

$$T\,(\text{in seconds}) = \frac{1}{F\,(\text{in hertz})}$$

**Fig. 8-1.** Relationship of wavelength, frequency, and period.

The *velocity* of a wave can be defined as the distance covered in a certain amount of time. This can be written

$$\text{velocity} = \frac{\text{distance}}{\text{time}} \text{ or } V = \frac{D}{T}$$

The velocity of a radio wave (or a light wave) in space is 186,000 miles (or 300,000,000 m) per second or 984 feet per miscrosecond.

If the period of a wave is defined as the time it takes the wave to travel a distance equal to its wavelength, then

$$\text{velocity} = \frac{\text{wavelength}}{\text{time}}$$

$$= \frac{\text{wavelength}}{(1/\text{frequency})}$$

$$= \text{frequency} \times \text{wavelength}$$

## 8-2. Time-Duration Measurements

An oscilloscope is the ideal tool for measuring time duration (and frequency) of voltages and currents. If the oscilloscope horizontal sweep is calibrated directly in relation to time, such as 5 ms/cm, the time duration of voltage waveforms may be measured directly on the screen without calculation. If the time duration of one complete cycle is measured, frequency can be calculated by simple division, since frequency is the reciprocal of the time duration of one cycle. If the oscilloscope is of the shop type in which the horizontal axis is not calibrated directly in relation to time, it is still possible to obtain accurate frequency and time measurements using Lissajous figures.

### 8-2.1. Time-duration Measurements with Laboratory Oscilloscopes

The horizontal sweep circuit of a laboratory oscilloscope usually has a selector control that provides a direct reading in relation to time. That is, each horizontal division on the oscilloscope screen has a definite relation to time at a given position of the horizontal sweep rate switch (such as milliseconds per centimeter). With such oscilloscopes the waveform can be displayed, and the time duration of the complete waveform (or any portion) can be measured directly as follows.

1. Connect the equipment as shown in Fig. 8-2.
2. Place the oscilloscope in operation as described in the instruction manual.
3. Set the vertical step-attenuator to a deflection factor that will allow the expected signal to be displayed without overdriving the vertical amplifier.
4. Connect the probe (if any) to the signal being measured.
5. Switch on the oscilloscope internal recurrent sweep. Set the horizontal sweep control to the fastest sweep rate that will display a convenient number of divisions between the time measurement points (Fig. 8-2).

### NOTE

On most oscilloscopes it is recommended that the extreme edges of the screen not be used for time-duration measurements, for there may be some nonlinearity at the beginning and end of the sweep.

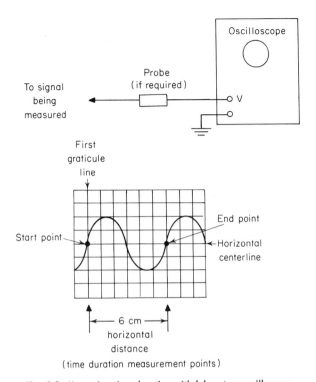

**Fig. 8-2.** Measuring time duration with laboratory oscilloscope.

6. Adjust the vertical position control to move the points between which the time measurement is made to the horizontal center line.

7. Adjust the horizontal position control to move the starting point of the time measurement area to the first graticule line.

8. Measure the horizontal distance between the time measurement points (Fig. 8-2).

### NOTE

If the horizontal sweep is provided with a variable control, make certain that it is off or in the calibrate position.

9. Multiply the distance measured in Step 8 by the setting of the horizontal sweep control. If sweep magnification is used, divide the answer by the multiplication factor.

10. As an example, assume that the distance between the time measure-

ment points is 6 cm (Fig. 8-2), the horizontal sweep control is set to 0.1 ms/cm, and there is no sweep magnification. Using the equation

$$\text{time duration} = \frac{\text{horizontal distance (cm)} \times \text{horizontal sweep setting}}{\text{magnification}}$$

substitute the given values

$$\text{time duration} = \frac{6 \times 0.1}{1} = 0.6 \text{ ms}$$

### 8-2.2. Time-duration Measurements with Shop Oscilloscopes

#### NOTE

The horizontal sweep circuit of most shop oscilloscopes is provided with controls that provide a direct reading in relation to frequency (whereas laboratory oscilloscopes read in relation to time). Usually, there are two controls—a step selector and a vernier. The sweep frequency of the horizontal trace is equal to the scale settings of the controls. Therefore, when a signal is applied to the vertical input and the horizontal sweep controls are adjusted until *one complete cycle* occupies the entire length of the trace, the vertical signal is equal in frequency to the horizontal sweep control scale setting. With the frequency established, time duration can be found using the equation, time $= 1/\text{frequency}$.

1. Connect the equipment as shown in Fig. 8-3.
2. Place the oscilloscope in operation as described in the instruction manual.
3. Set the vertical step-attenuator to a deflection factor that will allow the expected signal to be displayed without overdriving the vertical amplifier.
4. Connect the probe (if any) to the signal being measured.
5. Switch on the oscilloscope internal recurrent sweep. Set the horizontal sweep controls (step and vernier) so that one complete cycle occupies the entire length of the trace (Fig. 8-3).
6. Read the unknown vertical signal frequency directly from the horizontal sweep frequency control settings.
7. As an example, assume that the step horizontal sweep control is set to the 10-kHz position and that the vernier sweep control indicates 5 (on a total scale of 10), indicating that the horizontal sweep frequency is 5 kHz. If one complete cycle of vertical signals occupies the entire length

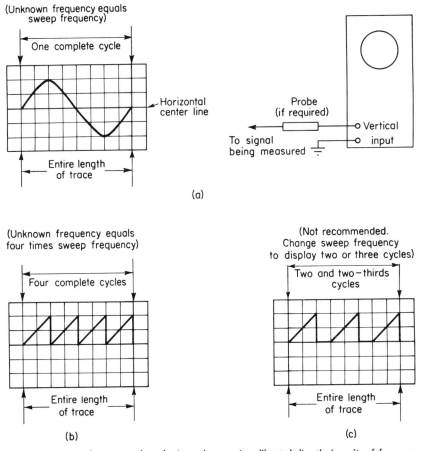

**Fig. 8-3.** Measuring frequency where horizontal sweep is calibrated directly in units of frequency.

of the trace, the vertical signal is also at a frequency of 5 kHz. This can be converted into a time duration using the equation

$$\text{time duration} = \frac{1}{\text{frequency}}$$

or

$$\frac{1}{5 \text{ kHz}} = 0.2 \text{ ms}$$

8. If it is not practical to display only one cycle on the trace, more than one cycle can be displayed, and the resultant horizontal sweep frequency indication multiplied by the number of cycles. However, certain precautions should be observed.

*First,* it is necessary to multiply the indicated sweep frequency by the number of cycles appearing on the trace. For example, if the indicated horizontal sweep frequency is 5 kHz and there are three cycles of vertical signal indicated, the true signal frequency is 15 kHz. In that case, the time duration would be $1 \div 15 \text{ kHz} = 0.066 \text{ ms}$.

*Second,* the exact number of cycles must occupy the *entire length* of the trace. As an example, if the indicated horizontal sweep frequency is 5 kHz and there are 3½ cycles of vertical signal indicated, the true signal frequency is 16.5 kHz. However, the exact percentage of the incomplete cycle (one-third) is quite difficult to determine. It is far simpler and more accurate to increase the horizontal sweep frequency until exactly three cycles appear or decrease the frequency until four cycles occupy the entire length of the trace.

## 8-3. Frequency Measurements with an Oscilloscope (Internal Sweep)

In the case of a laboratory oscilloscope, the frequency measurement of a periodically recurrent waveform is essentially the same as a time-duration measurement (Section 8-2.1), except that an additional calculation must be made. In effect, a time-duration measurement is made, and the time duration is divided into 1, or unity. Since frequency of a signal is the reciprocal of one cycle, frequency $= 1/(\text{time duration})$.

In the case of a shop oscilloscope, where the horizontal sweep controls are read in terms of frequency (Section 8-2.2), it is only necessary to adjust the horizontal sweep controls until one complete cycle occupies the entire length of the sweep. Under these conditions the vertical (or unknown) signal is equal in frequency to the horizontal sweep control settings (which can be read directly in terms of frequency).

## 8-4. Frequency Measurements with Lissajous Figures

Lissajous figures or patterns can be used with almost any oscilloscope (shop or laboratory) and will provide accurate frequency measurements. It must, however, be possible to apply an external signal to the horizontal amplifier with the internal sweep disabled. Also, an accurately calibrated variable-frequency signal source must be available to provide a standard frequency.

The use of Lissajous figures to measure frequency involves comparing a signal of unknown frequency (usually applied to the vertical amplifier) against a standard signal of known frequency (usually applied to the horizontal amplifier). The standard frequency is then adjusted until the pattern appears as a circle or ellipse, indicating that both signals are at the same frequency. Where it is not possible to adjust the standard signal frequency to the exact frequency of the unknown signal, the standard is adjusted to a multiple or submultiple. The pattern then appears as a number of stationary loops. The ratio of horizontal loops to vertical loops provides a measure of frequency.

1. Connect the equipment as shown in Fig. 8-4.

2. Place the oscilloscope in operation as described in the instruction manual.

3. Set the vertical step-attenuator to a deflection factor that will allow the expected signal to be displayed without overdriving the vertical amplifier.

4. Switch off the oscilloscope internal recurrent sweep.

5. Set the gain controls (horizontal and vertical) to spread the pattern over as much of the screen as desired.

6. Set the position controls (horizontal and vertical) until the pattern is centered on the screen.

7. Adjust the standard signal frequency until the pattern stands still. This indicates that the standard signal is at the same frequency as the unknown frequency (if the pattern is a circle or ellipse) or that the standard signal is at a multiple or submultiple of the unknown frequency (if the pattern is composed of stationary loops). If the pattern is still moving (usually spinning) it indicates that the standard signal is not at the same frequency (or multiple) of the unknown frequency. The pattern must be stationary before the frequency can be determined.

8. Note the standard signal frequency. Using this frequency as a basis, observe the Lissajous pattern and compare it against those shown in Fig. 8-4 to determine the unknown frequency.

## NOTE

If both signals are sinusoidal and are at the same frequency, the pattern will be a circle (or an ellipse, when the two signals are not exactly in phase) as shown in Fig. 8-4a.

If the standard signal (horizontal) is a multiple of the unknown signal (vertical) the pattern will show more horizontal loops than vertical loops (Fig. 8-4b). For example, if the standard signal frequency is three times that of the unknown signal frequency, there will be three horizontal loops and one vertical loop. If the standard-signal frequency were 300 Hz, the

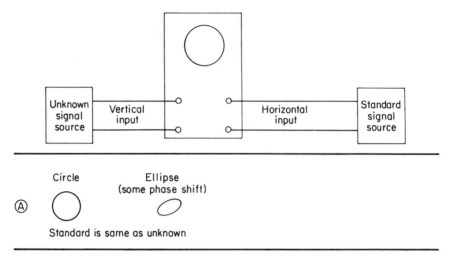

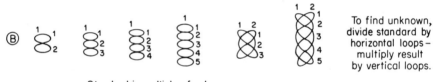

Fig. 8-4. Measuring frequency with Lissajous patterns.

unknown signal frequency would be 100 Hz. Or if there are two vertical loops and three horizontal loops, the unknown signal frequency would be two-thirds of the standard signal frequency.

If the standard signal (horizontal) is a submultiple of the unknown signal (vertical), the pattern will show more vertical loops than horizontal loops (Fig. 8-4c). For example, if the standard signal frequency is one-fourth that of the unknown signal frequency, there will be four vertical loops and one horizontal loop. If the standard signal frequency were 100 Hz, the unknown signal frequency would be 400 Hz.

9. If two signals are to be matched without regard to frequency, it is necessary to adjust only one frequency until the circle (or ellipse) pattern is obtained.

## NOTE

It is recommended that the circle pattern be used for all frequency measurements whenever possible. If this is not practical, use the minimum number of loops possible. Note too, that the use of Lissajous patterns in actual practice is quite difficult and requires considerable skill and practice to make accurate measurements.

## 8-5. Frequency Measurements with Modulated Ring Patterns

In some cases, it may be difficult to use Lissajous patterns, especially when there are many loops to be counted. An alternate method is use of a modulated ring. With this method, the display appears as a wheel or gear with a number of teeth. As with Lissajous figures, the modulated ring method requires that an external signal be applied with the internal sweep disabled and that the external signal be an accurately calibrated, variable source to provide a standard frequency.

Using the modulated ring method involves comparing a signal of unknown frequency (usually applied to the horizontal amplifier) against the standard signal of known frequency (usually applied through a phase shift network to the vertical amplifier). The standard frequency is then adjusted until the pattern stands still and appears as a circle (or ellipse) with a number of teeth or spikes. The number of teeth indicates the frequency of the unknown signal when multiplied by the known standard frequency.

It is essential that the unknown frequency be higher than the standard frequency to obtain a proper pattern. Also, it must be possible to increase the amplitude of the known signal to above that of the unknown signal to prevent distortion of the pattern.

1. Connect the equipment as shown in Fig. 8-5.

## NOTE

The phase shift required to produce a circle (or ellipse) pattern is composed of variable resistance $R_1$ and fixed capacitor $C_1$. To obtain the correct voltage to produce a good circle, the resistance of $R_1$

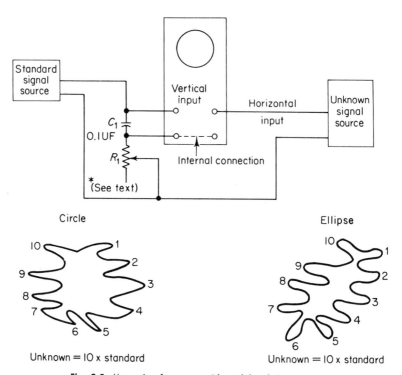

Fig. 8-5. Measuring frequency with modulated ring pattern.

should be equal to the reactance of $C_1$ at the operating frequency of the known signal source. By making $R_1$ variable, it is possible to match the change in $C_1$ reactance over a wide range of standard signal frequencies.

2. Place the oscilloscope in operation as described in the instruction manual.

3. Set the step-attenuators to deflection factors that will allow the expected signals to be displayed without overdriving the amplifier.

4. Switch off the oscilloscope internal recurrent sweep.

5. Temporarily remove the unknown signal source.

6. Set the position controls (horizontal and vertical) until the pattern is centered on the screen.

7. Adjust $R_1$ for a ring pattern.

8. Set the gain controls (horizontal and vertical) to spread the pattern over as much of the screen as desired.

9. Switch on the unknown signal source and note that the ring pattern

becomes modulated (teeth or spikes appear). It may be necessary to read just the horizontal gain to produce a readable pattern.

10. Adjust the standard signal frequency until the pattern stands still. This indicates that the unknown signal is at a multiple of the known signal frequency. The pattern must be stationary before the frequency can be determined.

11. Note the standard signal frequency. Using this frequency as a basis, observe the number of teeth appearing on the ring and determine the unknown frequency.

12. For example, assume that the standard signal frequency is 1 kHz and that there are seven teeth appearing on the ring, 1 kHz $\times$ 7 = 7-kHz unknown frequency.

## 8-6. Frequency Measurements with Broken Ring Patterns

The broken ring pattern is similar to the modulated ring pattern described in Section 8-5. Both are alternate methods to the use of Lissajous figures for frequency measurement. With the broken ring method the display appears as a ring broken into segments. As with the modulated ring method, the broken ring method requires that an external signal be applied with internal sweep disabled and that the external signal be an accurately calibrated variable source to provide standard frequency. An additional requirement of the broken ring method is that the oscilloscope must be capable of $Z$-axis (intensity) modulation.

Use of the broken ring method involves comparing a signal of unknown frequency (applied to the $Z$ axis) against the standard signal of known frequency (applied to the vertical and horizontal amplifiers through a phase-shift network). The standard frequency is then adjusted until the pattern stands still and appears as a circle (or ellipse) with a number of bright traces and blanks. The bright traces are made by the positive half cycles of the unknown signal applied to the $Z$ axis, whereas the blanks are made by the negative half cycles.

The number of traces (or blanks, whichever are easier to read) multiplied by the known standard frequency indicates the frequency of the unknown signal. It is essential that the unknown frequency be higher than the standard frequency to obtain a proper pattern.

The broken-ring method is superior to the modulated-ring method in that it is usually easier to distinguish blanks or traces than teeth in the ring pattern. Except for this the accuracy of both methods is the same.

1. Connect the equipment as shown in Fig. 8-6.

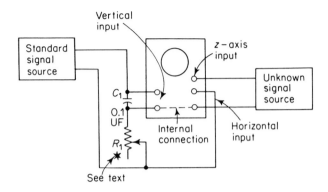

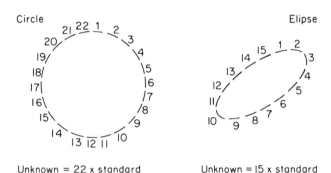

Unknown = 22 x standard             Unknown = 15 x standard

**Fig. 8-6.** Measuring frequency with broken ring pattern.

## NOTE

The phase shift required to produce the circle (or ellipse) pattern is composed of variable resistance $R_1$ and fixed capacitor $C_1$. To obtain the correct voltage to produce a good circle, the resistance of $R_1$ should equal the reactance of $C_1$ at the operating frequency of the known signal source. By making $R_1$ variable, it is possible to match the change in $C_1$ reactance over a wide range of standard signal frequencies.

2. Place the oscilloscope in operation as described in the instruction manual.

3. Set the step-attenuators to deflection factors that will allow the expected signals to be displayed without overdriving the amplifiers.

4. Switch off the oscilloscope internal recurrent sweep.

5. Temporarily remove the unknown signal source.

6. Set the position controls (horizontal and vertical) until the pattern is centered on the screen.

7. Adjust $R_1$ for a ring pattern on the screen.

8. Set the gain controls (horizontal and vertical) to spread the pattern over as much of the screen as desired.

9. Switch on the unknown signal source and note that the ring pattern is broken into segments.

10. Adjust the standard signal frequency until the pattern stands still. This indicates that the unknown signal is at a multiple of the known signal frequency. The pattern must be stationary before the frequency can be determined.

11. Note the standard signal frequency. Using this frequency as a basis, observe the number of traces (or blanks, whichever is more convenient to read) appearing on the ring and determine the unknown frequency.

12. For example, assume that the standard signal frequency is 1 kHz and there are 13 traces (or blanks) appearing on the ring. 1 kHz $\times$ 13 = 13-kHz unknown frequency.

## 8-7. Frequency Measurements with Broken Line Patterns

The broken line pattern is similar to the broken ring pattern described in Section 8-7. Both are alternate methods to the use of Lissajous figures for frequency measurement. With the broken line method, the display appears as a straight horizontal line (the vertical deflection is not used) broken into segments. As with the broken ring method, the broken line method requires that an external signal be applied with the internal sweep disabled and that the external signal be an accurately calibrated variable source to provide a standard frequency.

The broken line method requires that the oscilloscope be capable of Z-axis (intensity) modulation but does not require a phase-shift network.

Use of the broken line method involves comparing a signal of known frequency (applied to the Z axis) against the standard signal of known frequency (applied to the horizontal amplifier). The standard frequency is then adjusted until the pattern stands still and appears as a straight line with a number of bright traces and blanks. The bright traces are made by the positive half cycles of the unknown signal applied to the Z axis, whereas the blanks are made by the negative half cycles.

The number of traces (or blanks, whichever is easier to read) multiplied by the known standard frequency indicates the frequency of the unknown signal. It is essential that the unknown frequency be higher than the standard frequency to obtain a proper pattern.

The broken line method is superior to the broken ring method because it is simpler and does not require a phase-shift network, but the broken line method does not permit as high a count as the broken ring method.

1. Connect the equipment as shown in Fig. 8-7.
2. Place the oscilloscope in operation as described in the instruction manual.
3. Set the horizontal step-attenuator to a deflection factor that will allow the expected signals to be displayed without overdriving the horizontal amplifier.
4. Switch off the oscilloscope internal recurrent sweep.
5. Temporarily remove the unknown signal source.
6. Set the position controls (horizontal and vertical) until the pattern is centered on the screen.
7. Set the horizontal gain control to spread the pattern over as much of the screen as desired.
8. Switch on the unknown signal source and note that the horizontal line is broken into segments.
9. Adjust the standard signal frequency until the pattern stands still. This indicates that the unknown signal is at a multiple of the known signal frequency. The pattern must be stationary before the frequency can be determined.
10. Note the standard signal frequency. Using this frequency as a basis, observe the number of traces (or blanks, whichever is more convenient to read) appearing on the horizontal line and determine the unknown frequency.
11. For example, assume that the standard signal frequency is 1 kHz

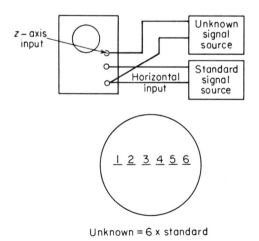

Unknown = 6 x standard

**Fig. 8-7.** Measuring frequency with broken line pattern.

and there are six traces (or blanks) appearing on the line. 1 kHz $\times$ 6 = 6-kHz unknown frequency.

## 8-8. Zero-beat Frequency Measurement

The basic components required for any form of zero-beat frequency measurement are a signal or voltage source of known accuracy that can be varied over the frequency range of the unknown signal or voltage and a detector that will show when the two signals are at the same frequency or some exact multiple of each other.

The Lissajous patterns as well as the modulater ring, broken ring, and broken line patterns described in Sections 8-4 through 8-7 are a form of zero-beat frequency measurement. In this case the oscilloscope is used as the detector, and the resultant patterns are used to indicate relationship of one signal frequency to the other.

When an oscilloscope is used as the detector, the signal source is usually an audio generator or low-frequency RF generator. When high-frequency RF signals are to be measured, an oscilloscope is not satisfactory as a detector since the passband of most oscilloscopes does not extend into the higher RF ranges. However, there are special-purpose laboratory oscilloscopes that have high-frequency amplifiers.

A radio receiver is a common detector for zero-beat frequency measurement. In fact, the term *zero beat* originated when radio receivers were used as frequency detectors. In operation, the two signals (known and unknown frequency) are applied to the radio receiver input. As the known signal frequency is adjusted close to that of the unknown signal (so that the difference in frequency is within the audio range) a tone, whistle, or "beat note" is heard on the receiver.

For example, assume that the unknown signal is at 5 MHz and the known signal frequency is adjusted to 5.001 MHz (or 10 kHz difference) so that the 10-kHz-difference signal can be heard. As the known signal frequency is brought closer to the unknown, say to 5.0001 (or 100-Hz difference), a tone is still heard. When the known signal is adjusted to exactly the same frequency as the unknown, there is no "difference" signal, and the tone can no longer be heard. In effect, the tone drops to "zero" and the two signals are at "zero beat."

### 8-8.1. Zero-beat Frequency Measurement with
#### Government Standards

A receiver can be used effectively for zero-beat measurements of frequency according to government standards. For example, an RF signal generator can be checked for proper calibration against the carrier frequency of the government standard stations of WWV (Fort Collins, Colorado) or WWVH (Honolulu). These stations make excellent standards for either shop or laboratory work since they operate over a wide frequency range and are highly accurate.

In the following procedure, the unknown signal frequency is variable, whereas the standard or known signal frequency is fixed. The unknown signal is varied until a zero beat is obtained. The accuracy of the unknown can then be verified against the known signal frequency. Note that the receiver frequency dial or indicator need not be highly accurate, but it is necessary to identify the stations correctly.

1. Connect the equipment as shown in Fig. 8-8. Operate the receiver and signal generator as described in their respective instruction manuals.

2. Tune the receiver to one (or more) of the government standard station frequencies.

### NOTE

For complete information on the WWV and WWVH schedules, refer to NBS (National Bureau of Standards) *Standard Frequency and Time Services* (Miscellaneous Publication 236) available from Superintendent of Documents, U.S. Government Printing Office, Washington D.C. 20402. The following is a summary of the WWV and WWVH schedules.

The NBS radio stations provide standard radio frequencies, standard audio frequencies, musical pitch, time intervals (1 second, 1 minute, 1 hour, etc.), time signals indicating Universal Time, Universal Time corrections (UT2), propagation forecasts, and geophysical alerts. The hourly broadcast schedules of WWV and WWVH are shown in Fig. 8-9.

Radio broadcasts from WWV on 2.5, 5, 10, 15, 20, and 25 MHz are

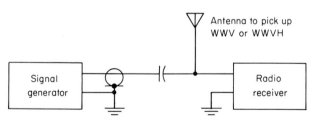

**Fig. 8-8.** Zero-beat frequency calibration of signal generator against standard frequency signals (WWV or WWVH).

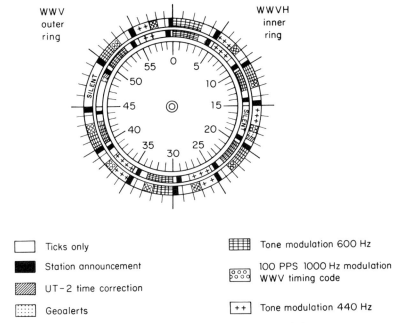

WWV outer ring

WWVH inner ring

SILENT    SILENT

55    0    5
50    10
45    15
40    20
35    25
30

| | Ticks only | | Tone modulation 600 Hz |
|---|---|---|---|
| ■ | Station announcement | | 100 PPS 1000 Hz modulation WWV timing code |
| ▨ | UT-2 time correction | | |
| ▤ | Geoalerts | ++ | Tone modulation 440 Hz |

**Fig. 8-9.** Summary of WWV and WWVH transmissions.

continuous night and day, except for silent periods of approximately 4 minutes beginning 45 minutes after each hour. Broadcast frequencies are held constant to within 5 parts in $10^{11}$. WWVH transmits on the four lower frequencies with 4-minute silent periods beginning 15 minutes after each hour. WWVB and WWVL broadcast continuously on 60 and 20 kHz, respectively, with a normal stability of two parts in $10^{11}$.

Standard audio frequencies of 440 and 600 Hz are broadcast on all WWV and WWVH frequencies. The tones are transmitted alternately in 5-minute intervals beginning with 600 Hz on the hour. The first tone transmission from WWV is 3 minutes long, all others are 2 minutes long. At WWVH, all tone transmissions are of 3-minute duration. WWVB and WWVL do not transmit standard audio tones.

For time signals, the audio frequencies are interrupted 3 minutes before each hour at WWV and 2 minutes before each hour at WWVH. The tones are restored on the hour and at 5-minute intervals throughout the hour.

Universal Time (referenced to the zero meridian at Greenwich, England) is announced in International Morse Code every 5 minutes from WWV and WWVH. At WWV, a voice announcement in Mountain Standard Time is made during the last half of each fifth minute during the hour.

At WWVH, a similar voice announcement of Hawaiian Standard Time is made during the first half of each fifth minute. WWVB transmits a special time code. WWVL does not transmit time signals.

Corrections to be applied to Universal Time signals are given in International Morse Code during the last half of the 19th minute from WWV and during the last half of the 49th minute from WWVH. The correction symbols consist of UT2, followed by AD (for add) or SU (for subtract), followed by a 3-digit number indicating the correction in milliseconds.

Propagation forecast announcements in Morse Code indicating ionospheric conditions and the quality of radio reception that can be expected within the next 6 hours are made from WWV at 0500, 1200 (1100 in summer), 1700, and 2300 UT. The forecast consists of a letter and a number. The letters *N, U,* and *W* indicate normal, unsettled, and disturbed conditions, respectively. Numbers 1 to 4 (disturbed) indicate useless, very poor, poor, and poor-to-fair; 5 (unsettled) is fair; 6 to 9 (normal) indicate fair-to-good, good, very good, and excellent, respectively.

A letter symbol indicating the current geophysical alert is broadcast in slow Morse Code from WWV during the first half of the 19th minute of each hour and from WWVH during the first half of each 49th minute. The geophysical alert identifies days in which outstanding solar or geophysical events are expected or have occurred during the preceding 24-hour period.

The geophysical announcement is identified by the letters GEO followed by the letter symbol repeated five times. The letter *M* means magnetic storm, and *N* is magnetic quiet, *C* is cosmic-ray event, *E* is no geoalert issued, *S* indicates the presence of solar activity, and *Q* is solar quiet, and *W* is stratospheric warning.

3. Set the signal generator to obtain a zero beat against the station carrier.

4. When a zero beat is heard in the receiver loudspeaker, the signal generator has an output frequency equal to that of the station carrier frequency.

## NOTE

As the signal generator is tuned to one-half (or other multiples) of that station frequency, a whistle or zero beat will be heard. This is because most signal generators have some harmonic output. However, the *fundamental* zero-beat note is the loudest. Good use can be made of harmonic beats, since one station of known frequency can be used to calibrate several generator bands. It should be noted that if the signal generator remains in the zero-beat condition over a long period of time (especially with the harmonic zero beats), the thermal stability of the generator is good.

5. If it is desired to calibrate an audio generator against the audio tones transmitted by WWV or WWVH, connect the equipment as shown in Fig. 8-10. Tune the receiver to one of the government standard station frequencies. Set the oscilloscope controls as necessary to display a Lissajous pattern. Under these conditions, the oscilloscope horizontal sweep will be obtained from the 440- or 600-Hz tone from WWV or WWVH, while vertical deflection will be obtained from the audio generator to be calibrated. Adjust the audio generator until the pattern stands still. This indicates that the audio generator is at the same frequency as the standard tone (if the pattern is a circle or ellipse) or that the audio generator is at a multiple or submultiple of the standard tone (if the pattern is composed of stationary loops). If the pattern is still moving (usually spinning), this indicates that the audio generator is not at the same frequency (or multiple) of the standard tone. The pattern must be stationary before the frequency can be determined.

6. Note the standard tone frequency (440 or 600 Hz). Using this frequency as a basis, observe the Lissajous pattern and compare it against those shown in Fig. 8-4 to determine the audio generator frequency. Refer to Section 8-4.

### 8-8.2. Zero-beat Frequency Measurement with a
### Crystal Oscillator

A signal generator (or other variable-frequency signal source) can be calibrated against a crystal oscillator. There are a number of easily constructed, simple crystal oscillator circuits (such as a Pierce oscillator). If

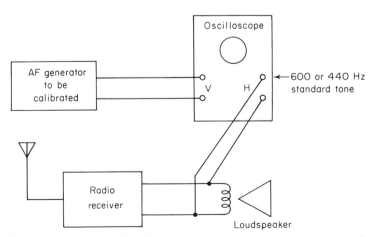

**Fig. 8-10.** Zero-beat frequency calibration of AF generator against standard tone signals (WWV or WWVH).

such an oscillator is used with a 1-MHz crystal, it will produce harmonics at 2 MHz, 3 MHz, etc. Also, if the generator is tuned to 100 kHz, its tenth harmonic will beat against the crystal fundamental. Therefore, the generator can be calibrated at frequencies much lower than those of the crystal fundamental.

1. Connect the equipment as shown in Fig. 8-11.

2. Tune the signal generator while listening for beat notes in the headphones. A zero beat indicates that the fundamental (or a harmonic) of the generator is at the same frequency as the fundamental (or harmonic) of the oscillator.

3. When both fundamentals are beat together, the beat note will be strongest. The beats will grow weaker as they are removed (in frequency) from the fundamental.

## 8-9. Measuring Resonant Frequency with Dip Adapters

The frequency of a resonant circuit can be found using the dip circuits described in Chapters 4 and 5. The following procedure is applicable to both series and parallel resonant circuits.

1. Couple the dip adapter (Fig. 4-18) to the circuit under test, using

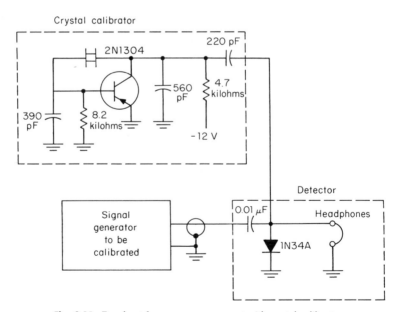

**Fig. 8-11.** Zero-beat frequency measurement with crystal calibrator.

the most convenient method of coupling (refer to Section 5-5 for coupling methods). De-energize the circuit.

2. Set the signal generator to its lowest frequency. Adjust the signal generator output amplitude control for a convenient reading on the adapter meter.

### NOTE

The signal generator should be set to a point just below the approximate frequency of the circuit under test. If the test-circuit frequency is not known, the generator should be set to its lowest frequency and then increased in frequency until a dip is found. Several harmonic dips may be found. The first dip encountered will be the fundamental (if the generator is set to its lowest frequency initially), and the resonant frequency can be read at once.

3. Tune the signal generator frequency for maximum dip and read the resonant frequency on the signal generator tuning dial.

4. For maximum accuracy, check the dip frequency from both high and low sides. A significant difference in frequency readout indicates overcoupling between the dip adapter circuit and the circuit under test. Move the dip adapter away from the test circuit until the dip indication is just visible. This amount of coupling should provide maximum accuracy. (If there is difficulty in finding a dip, overcouple the adapter until a dip is found, then loosen the coupling and make a final check of frequency.)

6. The adapter meter indication may rise or fall gradually as the signal generator is tuned in frequency. Or, there may be an abrupt change in meter indication when the signal generator range or frequency band is changed. Both of these conditions are a result of a change in signal generator output and must not be confused with a dip indication. Check if there is a corresponding change in the signal generator output meter. (Most laboratory signal generators are provided with an output meter.)

### NOTE

A gradual rise or fall in generator output can be put to good use. The dip will be more pronounced when it is approached from the direction that causes the meter reading to *rise*.

7. If there is doubt as to whether the adapter is measuring the resonant frequency of the desired circuit or some nearby circuit, ground the circuit. If there is no change in the adapter dip reading, the resonance of another circuit is being measured.

8. The area surrounding the circuit being measured should be free of wiring scraps, solder drippings, etc., as the resonant circuit can be affected by them (especially at higher frequencies), resulting in inaccurate frequency readings. Keep fingers and hands as far away as possible from the adapter coil (to avoid adding body capacitance to the circuit under test).

9. All other factors being equal, the nature of a dip indication provides an approximate indication of the test circuit's Q. Generally, a sharp dip indicates a high Q, while a broad dip indicates a low Q.

10. A dip adapter can also be used to preset a resonant circuit to some particular frequency. The procedure is essentially the same as that for finding the resonant frequency (Steps 1 through 9), except that the signal generator is set to the desired frequency, then the resonant circuit is tuned for a dip indication. Once the circuit has been tuned (for maximum dip), set the signal generator to its lowest frequency, then gradually increase the generator frequency, checking for resonant-frequency dip and harmonic dips. The largest dip should be at the resonant frequency.

## 8-10. Measuring Resonant Frequency of LC Circuits with a Meter

A meter can be used in conjunction with a signal generator (audio or RF) to find the resonant frequency of either series or parallel LC circuits (such as tank circuits, filters, etc.). The generator must be capable of producing a signal at the resonant frequency (not a harmonic) of the circuit, and the meter must be capable of measuring the frequency. If the resonant frequency is beyond the normal range of the meter, an RF probe must be used.

1. Connect the equipment as shown in Fig. 8-12. Use the connections of Fig. 8-12a for a parallel resonant circuit and the connections of Fig. 8-12b for a series resonant circuit.

2. Adjust the generator output until a convenient mid-scale indication is obtained on the meter. Use an unmodulated signal output from the generator.

3. Starting at a frequency well below the lowest possible frequency of the circuit under test, slowly increase the generator output frequency. If there is no way to judge the approximate resonant frequency, use the lowest generator frequency.

4. If the circuit being tested is parallel resonant, watch the meter for a maximum or peak indication.

5. If the circuit being tested is series resonant, watch the meter for a minimum or dip indication.

6. The resonant frequency of the circuit under test is the one at which

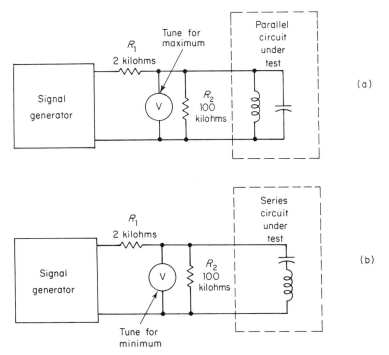

**Fig. 8-12.** Measuring resonant frequency of LC circuits.

there is a maximum (for parallel) or minimum (for series) indication on the meter.

7. There may be peak or dip indications at harmonics of the resonant frequency. Therefore, the test is most efficient when the approximate resonant frequency is known.

8. The value of load resistor $R_2$ is not critical. The load is shunted across the LC circuit to flatten or broaden the resonant response. Thus, the voltage maximum or minimum is approached more slowly. A suitable trial value for $R_2$ is 100,000 ohms. A lower value of $R_2$ will sharpen the resonant response, while a higher value will flatten the curve.

## 8-11. Sweep Frequency Technique for Measuring Resonant Frequency of Tuning Devices

The following basic procedure can be used to tune (or to check the resonant frequency) of any resonant circuit. The procedure is particularly effective with devices requiring transmission lines (such as an antenna). Often, antennas used with transmission lines are tuned by an external device. As an example of this, TV antennas are often tuned to cover a specific frequency by means of a stub. The sweep generator/oscilloscope combination described in Section 6-6.1 can be used to measure the resonant frequency of a tuning device.

1. Connect the equipment as shown in Fig. 8-13.

2. Place the oscilloscope in operation as described in the instruction manual. Switch off the internal recurrent sweep. Set the oscilloscope sweep selector and sync selector to external.

3. Place the sweep generator in operation as described in the instruction manual. Switch the sweep generator blanking control on or off as desired. Tune the sweep generator to the frequency at which the device is supposed to be resonant. Adjust the sweep width to cover a wide range of frequencies, but not so wide that the nearest harmonic is covered. If harmonics are covered, their indications on the trace may prove confusing.

4. Note the point on the trace at which the dip occurs. The dip indicates the frequency of the resonant circuit.

5. The procedure can be reversed to adjust the tuning device to a given frequency if desired.

6. For greatest accuracy, adjust the marker generator until the marker pip is aligned at the center of the trace dip. The exact resonant frequency can then be read from the marker generator frequency dial.

### NOTE

The dip characteristics indicate the approximate sharpness or Q of the resonant device. A broad dip indicates a low Q; a sharp dip indicates a high Q.

## 8-12. Measuring Frequency Response

The frequency response of an amplifier, filter, or any circuit with band-pass characteristics (receiver front end, overall receiver circuits, etc.) can be measured with a signal generator (audio or RF, as applicable) and a meter or oscilloscope. When a meter is used, the signal generator is tuned to various frequencies, and the resultant circuit output response is measured at each frequency. The results are often plotted in the form of a graph or response curve. With the oscilloscope method, the signal source is usually a sweep generator, permitting the entire frequency range to be measured (and displayed) simultaneously.

The following paragraphs describe the basic method of measuring frequency response of circuits, using both a meter and an oscilloscope.

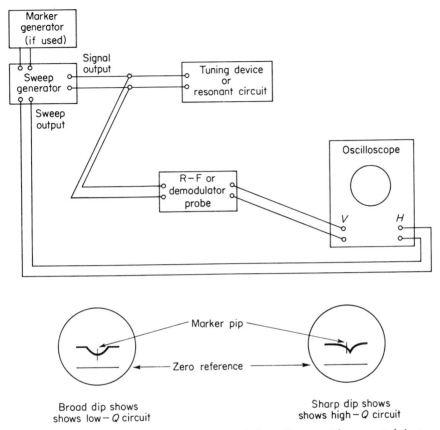

**Fig. 8-13.** Measuring resonant frequency of tuning devices with resonant frequency technique.

Also described are procedures for measuring frequency response of specific circuits such as filters.

### 8-12.1. Measuring Frequency Response with a Meter

The basic method is to apply a *constant amplitude* signal while monitoring the circuit output with the meter. The input signal is varied in frequency (but not amplitude) across the entire operating range of the circuit. Usually, this is from 20 Hz to 20 kHz for audio amplifiers, although some hi-fi and industrial amplifier manufacturers specify a response up to 100 kHz (or higher). The voltage output at various frequencies across the range is plotted on a graph similar to that shown in Fig. 8-14.

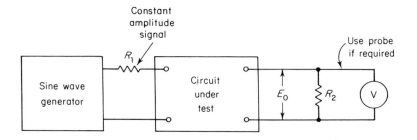

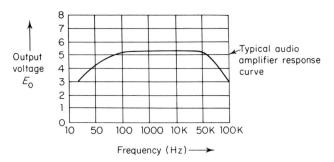

**Fig. 8-14.** Measuring frequency response with a meter.

1. Connect the equipment as shown in Fig. 8-14.

2. Set the meter to measure ac or "output," unless a probe is used.

3. Place the generator in operation. Set the generator output frequency to the lowest point specified in the manufacturer's data. (Use a low frequency of approximately 20 Hz for audio circuits in the absence of a specified low limit.)

4. Connect the meter to measure the output of the generator as it appears across the input of the first stage.

### NOTE

Resistors $R_1$ and $R_2$ are included since many test specifications for circuits require that the input and output be terminated in their respective impedances for proper voltage measurement. $R_1$ and $R_2$ may be omitted unless required by specifications.

5. Set the generator output level to the value recommended in the circuit manufacturer's data. In the absence of a specification set the generator output to an arbitrary value. A simple method of determining a

satisfactory input level is to monitor the circuit output with the meter and increase the generator output at the circuit center frequency until the circuit is overdriven. This point will be indicated when further increases in generator output do not cause further increases in meter reading. Set the generator output just below this point. Then return the meter to monitor the generator voltage (at circuit input) and measure the voltage. Keep the generator at this voltage throughout the test.

6. Set the circuit operating controls (if any) to the normal operating point or at the particular setting specified in the manufacturer's test data.

7. Record the circuit output voltages on the graph.

8. Without changing the generator output amplitude, increase the generator frequency by 100 Hz or as specified in the manufacturer's data. Record the new circuit output voltage on the graph. Repeat this process, checking and recording the circuit output voltage at each of the check points in order to obtain a frequency response curve. With a typical amplifier circuit, the curve will resemble that of Fig. 8-14, with a flat portion across the center and a roll-off at each end. Some amplifier circuits are designed to provide a high-frequency boost (where the high end of the curve increases in amplitude) or low-frequency boost (where the low end shows an amplitude increase). The manufacturer's data must be consulted for this information.

9. Note that generator output may vary with changes in frequency, a fact often overlooked in making a frequency-response test of any circuit. Even precision laboratory generators can vary in output with changes in frequency, which could result in considerable error. Therefore, it is recommended that the generator output be monitored after each change in frequency. Then, if necessary, the generator output amplitude can be reset to the correct value established in Step 5. Within extremes, it is more important that the generator output amplitude remain *constant* rather than at some specific value when making a frequency-response check.

10. Repeat the frequency-response check with the circuit operating controls (if any) at each of their positions or as specified in the manufacturer's data.

### 8-12.2. Measuring Frequency Response with an Oscilloscope

An oscilloscope can be used in place of a meter to measure frequency response. In this application, the oscilloscope is used as an audio or RF voltmeter. The procedure is essentially the same as that described in Section 8-12.1. However, an oscilloscope can be used more effectively when operated with a sweep generator, as described in Section 6-6. This technique permits the frequency response to be measured over the complete range simultaneously. The response characteristics of a circuit can be checked using the sweep generator/oscilloscope combination, provided that the sweep generator is capable of sweeping over the frequency range. If maximum accuracy is desired, a marker generator must also be used to measure exact frequency.

The following procedure can be used to measure frequency response of several circuits (filters, front end, IF, overall receiver response, amplifiers, etc.). Refer to Section 6-6 for information on the basic sweep technique.

1. Connect the equipment as shown in Fig. 8-15.

### NOTE

Resistors $R_1$ and $R_2$ are included since many test specifications require that the input and output be terminated in their respective impedances. $R_1$ and $R_2$ may be omitted unless required by specification.

2. Place the oscilloscope in operation as described in the instruction manual. Switch off internal recurrent sweep. Set the oscilloscope sweep selector and sync selector to external.

3. Place the sweep generator in operation as described in the instruction manual. Switch the sweep generator blanking control on or off as desired. Adjust the sweep generator to cover the complete frequency range or that portion of the range that would affect circuit operation.

4. Check the circuit response curve appearing on the oscilloscope screen against Fig. 8-15 or against the circuit specifications. Typical filter, amplifier, and receiver response curves are shown in Fig. 8-15.

5. If it is desired to find the exact frequencies at which circuit response occurs, the marker generator can be adjusted until the marker pip is aligned at the point of interest. The frequency or band of frequencies can be read from the marker generator frequency dial.

6. The amplitude of any point on the response curve can be measured directly on the oscilloscope (assuming that the vertical system is voltage calibrated).

## 8-13. Frequency Bridges

Precision frequency values are often measured by means of bridge circuits. Although there are many basic frequency bridge circuits in use, those shown in Figs. 8-16 through 8-25 are the best known. Operation of these circuits is similar to that of the resistance bridges described in Section

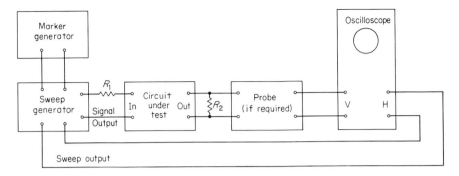

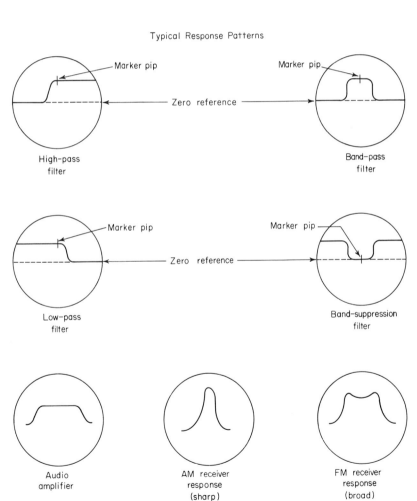

Typical Response Patterns

**Fig. 8-15.** Measuring frequency response with an oscilloscope using sweep technique.

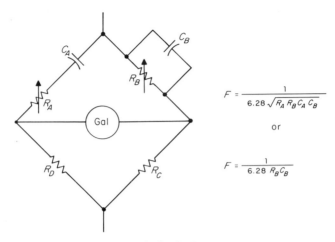

$$F = \frac{1}{6.28 \sqrt{R_A R_B C_A C_B}}$$

or

$$F = \frac{1}{6.28 \, R_B C_B}$$

**Fig. 8-16.** Basic Wien bridge for frequency measurement.

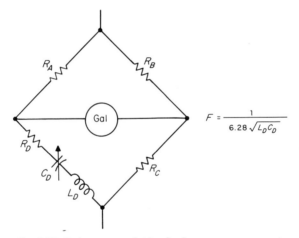

$$F = \frac{1}{6.28 \sqrt{L_D C_D}}$$

**Fig. 8-17.** Basic resonance bridge for frequency measurement.

3-7. The reasons for using commercial bridge circuits, together with their related instruction manuals, are discussed in Section 4-3.

## 8-14. Phase Angle Relationships

In alternating currents, a pair of currents, a pair of voltages, or the voltage and current need not be in step even though they exist in the same circuit. One current may lead the other, one voltage may lag the other,

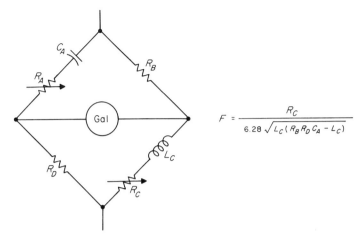

$$F = \frac{R_C}{6.28\sqrt{L_C(R_B R_D C_A - L_C)}}$$

**Fig. 8-18.** Basic Hay bridge for frequency measurement.

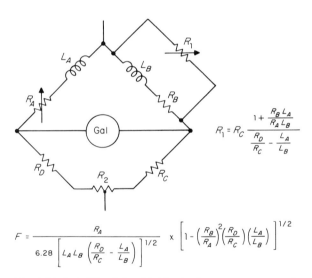

$$R_1 = R_C \frac{1 + \frac{R_B L_A}{R_A L_B}}{\frac{R_D}{R_C} - \frac{L_A}{L_B}}$$

$$F = \frac{R_A}{6.28\left[L_A L_B \left(\frac{R_D}{R_C} - \frac{L_A}{L_B}\right)\right]^{1/2}} \times \left[1 - \left(\frac{R_B}{R_A}\right)^2\left(\frac{R_D}{R_C}\right)\left(\frac{L_A}{L_B}\right)\right]^{1/2}$$

**Fig. 8-19.** Basic Wien-Dolezak bridge for frequency measurement.

or the voltage may lag the current. The amount of lead or lag is termed *phase angle* and is expressed by the Greek letter theta, $\theta$.

Since most alternating currents are sine waves, they have periodic waveforms where the lead or lag can be measured along the $X$ axis. Out-of-phase voltages (or currents) can be added vectorially to produce a resultant voltage (or current).

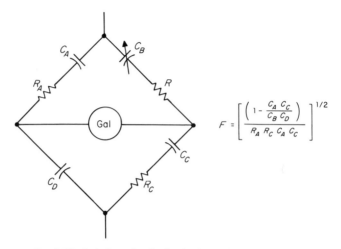

**Fig. 8-20.** Basic Sacerdote bridge for frequency measurement.

$$F = \left[ \frac{\left(1 - \frac{C_A \, C_C}{C_B \, C_D}\right)}{R_A \, R_C \, C_A \, C_C} \right]^{1/2}$$

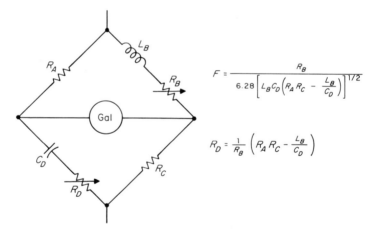

**Fig. 8-21.** Basic Hay (modified) bridge for frequency measurement.

$$F = \frac{R_B}{6.28 \left[ L_B C_D \left( R_A R_C - \frac{L_B}{C_D} \right) \right]^{1/2}}$$

$$R_D = \frac{1}{R_B} \left( R_A R_C - \frac{L_B}{C_D} \right)$$

In a *pure resistive* circuit, there will be no lag or lead between the current and voltage (*in phase*).

In a *pure inductive* circuit, the current will lag the voltage by 90°.

In a *pure capacitive* circuit, the current will lead the voltage by 90°.

These relationships are shown in Fig. 8-26.

Most a-c circuits are not purely resistive, inductive, or capacitive. Instead, they are combinations of all three factors. The ratio of reactance

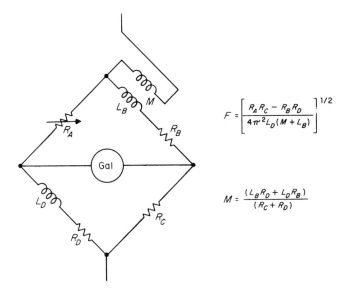

$$F = \left[ \frac{R_A R_C - R_B R_D}{4\pi^2 L_D (M + L_B)} \right]^{1/2}$$

$$M = \frac{(L_B R_D + L_D R_B)}{(R_C + R_D)}$$

Fig. 8-22. Basic Butterworth bridge for frequency measurement.

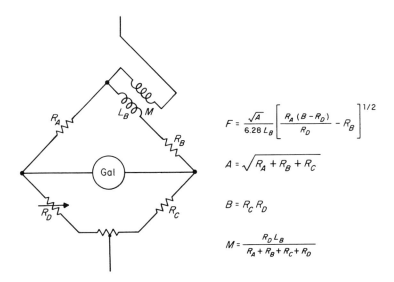

$$F = \frac{\sqrt{A}}{6.28 L_B} \left[ \frac{R_A (B - R_D)}{R_D} - R_B \right]^{1/2}$$

$$A = \sqrt{R_A + R_B + R_C}$$

$$B = R_C R_D$$

$$M = \frac{R_D L_B}{R_A + R_B + R_C + R_D}$$

Fig. 8-23. Basic Hughes-Campbell bridge for frequency measurement.

and resistance is indicative of the size of the phase angle. This may be
calculated by means of trigonometry. The relationship is shown in Fig.
8-27.

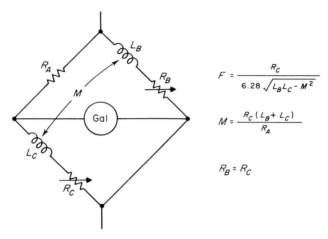

$$F = \frac{R_C}{6.28 \sqrt{L_B L_C - M^2}}$$

$$M = \frac{R_C (L_B + L_C)}{R_A}$$

$$R_B = R_C$$

**Fig. 8-24.** Basic Kurokawa bridge for frequency measurement.

If trigonometry is used, the altitude (reactance) is divided by the base (resistance) to produce the tangent of the phase angle. Table 8-1 lists the ratios (tangents) for phase angles from 0° to 89°.

**TABLE 8-1**

Phase Angle Versus Ratio of Reactance Divided by Resistance

| Phase Angle | Ratio | Phase Angle | Ratio | Phase Angle | Ratio | Phase Angle | Ratio | Phase Angle | Ratio |
|---|---|---|---|---|---|---|---|---|---|
| 0 | 0.0000 | 20 | 0.3640 | 40 | 0.8391 | 60 | 1.7321 | 80 | 5.6713 |
| 1 | 0.0175 | 21 | 0.3839 | 41 | 0.8693 | 61 | 1.8040 | 81 | 6.3138 |
| 2 | 0.0349 | 22 | 0.4040 | 42 | 0.9004 | 62 | 1.8870 | 82 | 7.1154 |
| 3 | 0.0524 | 23 | 0.4245 | 43 | 0.9325 | 63 | 1.9626 | 83 | 8.1443 |
| 4 | 0.0699 | 24 | 0.4452 | 44 | 0.9657 | 64 | 2.0503 | 84 | 9.5144 |
| 5 | 0.0875 | 25 | 0.4663 | 45 | 1.0000 | 65 | 2.1445 | 85 | 11.4301 |
| 6 | 0.1051 | 26 | 0.4877 | 46 | 1.0355 | 66 | 2.2460 | 86 | 14.3007 |
| 7 | 0.1228 | 27 | 0.5095 | 47 | 1.0724 | 67 | 2.3559 | 87 | 19.0811 |
| 8 | 0.1405 | 28 | 0.5317 | 48 | 1.1106 | 68 | 2.4751 | 88 | 28.6363 |
| 9 | 0.1584 | 29 | 0.5543 | 49 | 1.1504 | 69 | 2.6051 | 89 | 57.2900 |
| 10 | 0.1763 | 30 | 0.5774 | 50 | 1.1918 | 70 | 2.7475 | | |
| 11 | 0.1944 | 31 | 0.6009 | 51 | 1.2349 | 71 | 2.9042 | | |
| 12 | 0.2126 | 32 | 0.6249 | 52 | 1.2799 | 72 | 3.0777 | | |
| 13 | 0.2309 | 33 | 0.6494 | 53 | 1.3270 | 73 | 3.2709 | | |
| 14 | 0.2493 | 34 | 0.6745 | 54 | 1.3764 | 74 | 3.4874 | | |
| 15 | 0.2679 | 35 | 0.7002 | 55 | 1.4281 | 75 | 3.7321 | | |
| 16 | 0.2967 | 36 | 0.7265 | 56 | 1.4826 | 76 | 4.0108 | | |
| 17 | 0.3057 | 37 | 0.7536 | 57 | 1.5399 | 77 | 4.3315 | | |
| 18 | 0.3249 | 38 | 0.7813 | 58 | 1.6003 | 78 | 4.7046 | | |
| 19 | 0.3443 | 39 | 0.8098 | 59 | 1.6643 | 79 | 5.1446 | | |

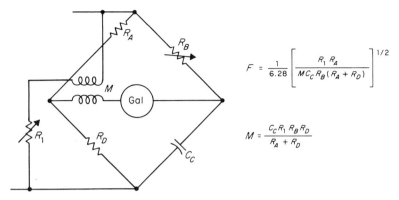

$$F = \frac{1}{6.28}\left[\frac{R_1\,R_A}{M C_C R_B(R_A + R_D)}\right]^{1/2}$$

$$M = \frac{C_C R_1\,R_B R_D}{R_A + R_D}$$

Fig. 8-25. Basic Schering and Engelhardt bridge for frequency measurement.

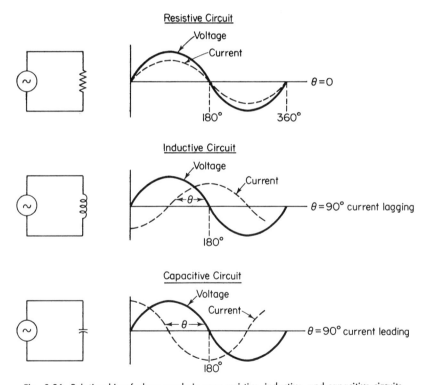

Fig. 8-26. Relationship of phase angle in pure resistive, inductive, and capacitive circuits.

For example, assume that the resistance was 30 ohms, the inductive reactance was 25 ohms, and the capacitive reactance was 5.5 ohms.

$$X_L - X_C = 25 - 5.5 = 19.5;$$

$$\frac{X}{R} = \frac{19.5}{30} = 0.65;$$

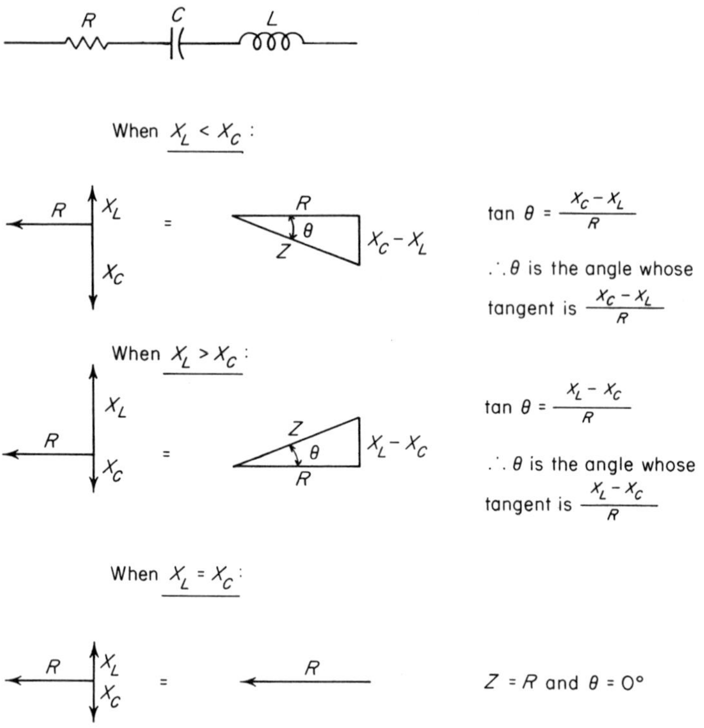

**Fig. 8-27.** Relationship of phase angle to resistance, capacitance, and reactance.

$\theta$ is an angle whose tangent is 0.65; therefore $\theta$ is approximately 33° with current lagging (Table 8-1).

Using either vectors or trigonometry, note that the smaller reactance (inductive or capacitive) must be subtracted from the larger before calculating the reactance/resistance ratio.

## 8-15. Phase Measurement

The oscilloscope is the ideal tool for phase measurement. There are two basic methods of phase measurement. The most convenient method requires a *dual-trace* oscilloscope or an electronic switching unit to provide a dual trace. If neither of these is available, it is still possible to provide accurate phase measurement up to about 100 kHz using the single-trace or *X-Y* method.

8-15.1. Phase Measurement Between Two
Voltages (Dual-trace)

The dual-trace method of phase measurement provides a high degree of accuracy at all frequencies but is especially useful at frequencies above 100 kHz where *X-Y* phase measurements may prove inaccurate owing to inherent internal phase shift of the oscilloscope.

The dual-trace method also has the advantage of measuring phase difference between signals of different amplitudes, frequency, and waveshape. The method can be applied directly to those oscilloscopes having a built-in dual-trace feature or to a conventional single-trace oscilloscope using an electronic switch or "chopper." Either way, the procedure is essentially one of displaying both traces on the oscilloscope screen simultaneously, measuring the distance (in scale divisions) between related points on the two traces, and then converting this distance into phase.

1. Connect the equipment as shown in Fig. 8-28.
2. Place the oscilloscope in operation as described in the instruction manual.

### NOTE

For the most accurate results, the cables connecting the two signals to the oscilloscope input should be of the same length and characteristics. At higher frequencies, a difference in cable length or characteristics could introduce a phase shift.

3. Set the step attenuators to deflection factors that will allow the expected signals to be displayed without overdriving the amplifiers.
4. Switch on the oscilloscope internal recurrent sweep.
5. Set the position controls (horizontal and vertical) until the pattern is centered on the screen.
6. Set the gain controls (horizontal and vertical) to spread the patterns over as much of the screen as desired.
7. Switch on the dual-trace function of the oscilloscope, or switch on the electronic "chopper."
8. Adjust the sweep controls until one cycle of the reference signal occupies exactly nine divisions (9 cm horizontally) of the screen.

### NOTE

Either of the two signals can be used as the reference signal unless otherwise specified by requirements of the particular test. It is usually simpler if the signal of the lowest frequency is used as the reference signal.

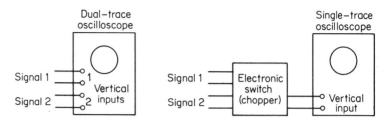

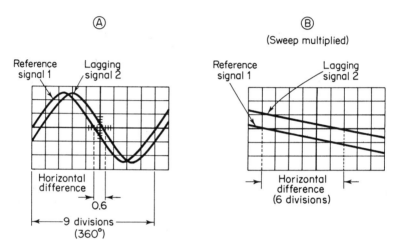

**Fig. 8-28.** Measuring phase difference with dual traces.

9. Determine the phase factor of the reference signal.

10. As an example, if 9 cm represents one complete cycle or 360°, then 1 cm represents 40° [360°/9 divisions (cm) = 40°/cm].

11. Measure the horizontal distance between corresponding points on the waveform. Multiply the measured distance (in centimeters) by 40° (phase factor) to obtain the exact amount of phase difference.

12. For example, assume a horizontal difference of 0.6 cm with a phase-angle factor of 40° as shown in Fig. 8-28. Multiply the horizontal difference by the phase-angle factor to find the phase difference (0.6 × 40 = 24° phase-angle difference between the two signals).

13. If the oscilloscope is provided with a sweep magnification control in which the sweep rate is increased by some fixed amount (5×, 10×, etc.) and only a portion of one cycle can be displayed, more accurate phase measurements can be made. In this case, the phase factor is determined as described in Step 9. Then the approximate phase difference

is determined as described in Step 11. Without changing any other controls, the sweep rate is increased (by the sweep magnification control or the sweep rate control), and a new horizontal distance measurement is made, as shown in Fig. 8-28b.

14. For example, if the sweep rate were increased 10 times, the adjusted phase factor would be $40°/10 = 4°/cm$. Figure 8-28b shows the same signal as used in Fig. 8-28a, but with the sweep rate set to $10\times$. With a horizontal difference of 6 cm, the phase difference would be $6 \times 4° = 24°$.

### 8-15.2. Phase Measurement Between Two Voltages (X-Y Method)

The $X$-$Y$ phase measurement method can be used to measure the phase difference between two sine-wave signals of the *same frequency*. This method measures signal frequencies with approximately 100 kHz more precision than the dual-trace method discussed in Section 8-15.1. Above this frequency, however, the inherent phase shift (or difference) between the horizontal and vertical systems makes accurate phase measurements difficult. Therefore, the $X$-$Y$ method should be limited to phase measurement of low-frequency signals and to signals of the same frequency.

In the $X$-$Y$ method, one of the sine-wave signals provides horizontal deflection ($X$), and the other provides the vertical deflection ($Y$). The phase angle between the two signals can be determined from the resulting Lissajous pattern.

1. Connect the equipment as shown in Fig. 8-29a.

### NOTE

Figure 8-29a shows the test connection necessary to find the inherent phase shift (if any) between the horizontal and vertical deflection systems of the oscilloscope. Even laboratory oscilloscopes with identical horizontal and vertical amplifiers will have some inherent phase shift, particularly at the higher frequencies. Therefore, all oscilloscopes should be checked and the inherent phase shift *recorded* before any phase measurements are made. Inherent phase shift should be checked periodically. If there is excessive phase shift (in relation to the signals to be measured), the oscilloscope should not be used. A possible exception exists when the signals to be measured are of sufficient amplitude to be applied directly to the oscilloscope deflection plates, bypassing the horizontal and vertical amplifiers.

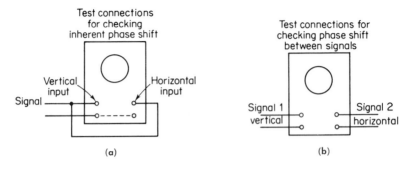

Test connections for checking inherent phase shift

Test connections for checking phase shift between signals

(a)             (b)

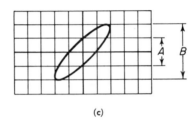

(c)

**Fig. 8-29.** Measuring phase difference with X-Y method.

2. Place the oscilloscope in operation as described in the instruction manual.

3. Set the step-attenuators to deflection factors that will allow the expected signals to be displayed without overdriving the amplifiers.

4. Switch off the oscilloscope internal recurrent sweep.

5. Set the gain controls (horizontal and vertical) to spread the pattern over as much of the screen as desired.

6. Set the position controls (horizontal and vertical) until the pattern is centered on the screen. Center the display in relation to the vertical graticule line. Measure distances A and B, as shown in Fig. 8-29c. Distance A is the vertical measurement between two points where the trace crosses the vertical center line. Distance B is the maximum vertical height of the display.

7. Divide A by B to obtain the sine of the phase angle between the two signals. The angle can then be obtained from Table 8-2. The resultant angle is the phase shift.

### NOTE

If the display appears as a diagonal straight line, the two amplifiers are either in phase (tilted upper right to lower left) or 180° out-of-phase (tilted from upper left to lower right). If the display is a circle,

**TABLE 8-2**

Table of Sines

| Sine | Angle | Sine | Angle | Sine | Angle | Sine | Angle | Sine | Angle |
|------|-------|------|-------|------|-------|------|-------|------|-------|
| 0.0000 | 0 | 0.3584 | 21 | 0.6691 | 42 | 0.8910 | 63 | 0.9945 | 84 |
| 0.0175 | 1 | 0.3746 | 22 | 0.6820 | 43 | 0.8988 | 64 | 0.9962 | 85 |
| 0.0349 | 2 | 0.3907 | 23 | 0.6947 | 44 | 0.9063 | 65 | 0.9976 | 86 |
| 0.0523 | 3 | 0.4067 | 24 | 0.7071 | 45 | 0.9135 | 66 | 0.9986 | 87 |
| 0.0689 | 4 | 0.4226 | 25 | 0.7193 | 46 | 0.9205 | 67 | 0.9994 | 88 |
| 0.0872 | 5 | 0.4384 | 26 | 0.7314 | 47 | 0.9272 | 68 | 0.9998 | 89 |
| 0.1045 | 6 | 0.4540 | 27 | 0.7431 | 48 | 0.9336 | 69 | 1.0000 | 90 |
| 0.1219 | 7 | 0.4695 | 28 | 0.7547 | 49 | 0.9397 | 70 | | |
| 0.1392 | 8 | 0.4848 | 29 | 0.7660 | 50 | 0.9455 | 71 | | |
| 0.1564 | 9 | 0.5000 | 30 | 0.7771 | 51 | 0.9511 | 72 | | |
| 0.1736 | 10 | 0.5150 | 31 | 0.7880 | 52 | 0.9563 | 73 | | |
| 0.1908 | 11 | 0.5299 | 32 | 0.7986 | 53 | 0.9613 | 74 | | |
| 0.2079 | 12 | 0.5446 | 33 | 0.8090 | 54 | 0.9659 | 75 | | |
| 0.2250 | 13 | 0.5592 | 34 | 0.8192 | 55 | 0.9703 | 76 | | |
| 0.2419 | 14 | 0.5736 | 35 | 0.8290 | 56 | 0.9744 | 77 | | |
| 0.2588 | 15 | 0.5878 | 36 | 0.8387 | 57 | 0.9781 | 78 | | |
| 0.2756 | 16 | 0.6018 | 37 | 0.8480 | 58 | 0.9816 | 79 | | |
| 0.2924 | 17 | 0.6157 | 38 | 0.8572 | 59 | 0.9848 | 80 | | |
| 0.3090 | 18 | 0.6293 | 39 | 0.8660 | 60 | 0.9877 | 81 | | |
| 0.3256 | 19 | 0.6428 | 40 | 0.8746 | 61 | 0.9903 | 82 | | |
| 0.3420 | 20 | 0.6561 | 41 | 0.8829 | 62 | 0.9925 | 83 | | |

the signals are 90° out-of-phase. Figure 8-30 shows the Lissajous displays produced between 0° and 360°. Notice that above a phase shift of 180°, the resultant display will be the same as at some lower frequency. Therefore, it may be difficult to tell whether the signal is leading or lagging. One way to find correct phase polarity (leading or lagging) is to introduce a small, known phase shift to one of the inputs. The proper angle may then be determined by noting the direction in which the pattern changes.

8. Once the inherent phase shift has been established, connect the equipment as shown in Fig. 8-29b. Repeat Steps 3 through 7 to find the phase angle between the two signals.

9. Subtract the inherent phase difference from the phase angle to find the true phase difference.

10. As an example, assume an inherent phase difference of 2°, with a display as shown in Fig. 8-29c, where $A$ is 2 cm and $B$ is 4 cm. Sine of phase angle $= A/B$, or $2/4$, or 0.5. From Table 8-2, 0.5 $= 30°$ phase angle. To adjust for the phase difference between $X$ and $Y$ amplifiers, subtract the inherent phase factor ($30° - 2° = 28°$ true phase difference).

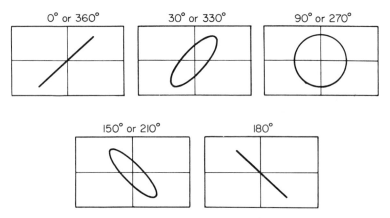

**Fig. 8-30.** Phase of typical X-Y displays.

### 8-15.3. Phase Measurement Between Voltage and Current

To measure the phase difference between a voltage and current applied to the same reactive load, a portion of the current can be passed through a fixed resistor, thus converting the current to a voltage. The phase of the resultant voltage (across the resistance) is then compared with the load voltage phase. A resistor, capable of the necessary wattage dissipation, is the only other component required for the procedure. Either the $X$-$Y$ method or the dual-trace method can be used for the actual phase comparison.

Figure 8-31 shows the test connections required for converting the current into a voltage and applying both voltages to the oscilloscope.

In Fig. 8-31a, the signal voltage $E_1$ is applied across the load and test resistor $R_1$. Voltage $E_1$ is also applied to one of the vertical inputs. The current-developed voltage $E_2$ appears across $R_1$ and is applied to the other vertical input.

In Fig. 8-31b, the signal voltage $E_1$ is applied across the load and test resistor $R_1$. Voltage $E_1$ is also applied to the electronic switch (chopper). The current-developed voltage $E_2$ appears across $R_1$ and is applied to the other electronic switch input.

In Fig. 8-31c, the signal voltage $E_1$ is applied across the load and test resistor $R_1$. Voltage $E_1$ is also applied to the vertical input. The current-developed voltage $E_2$ appears across $R_1$ and is applied to the horizontal input.

Once the test connections have been made, the phase difference between voltage and current can be determined by the procedures of Section 8-15.1 (dual-trace), or Section 8-15.2 ($X$-$Y$ method), whichever applies.

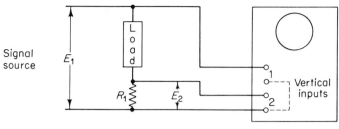

(a) Dual trace method (dual trace oscilloscope)

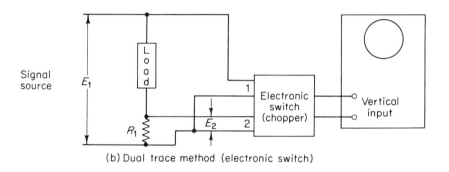

(b) Dual trace method (electronic switch)

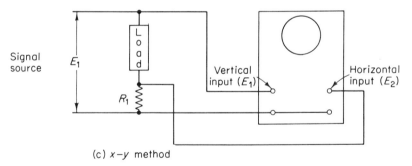

(c) x–y method

**Fig. 8-31.** Phase measurement between voltage and current.

## NOTE

The actual resistance value of $R_1$ is not critical, but it should be low in comparison to the resistance value of the load. Usually 1 to 10 ohms is adequate to develop sufficient voltage for measurement. The wattage of the resistor $R_1$ must be at least double the square of the maximum current (in amperes). For example, if the maximum

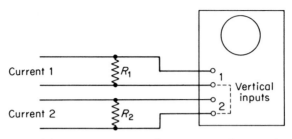

(a) Dual trace method (dual trace oscilloscope)

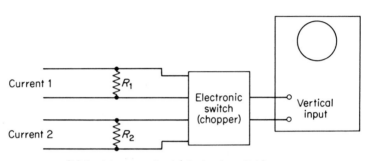

(b) Dual trace method (electronic switch)

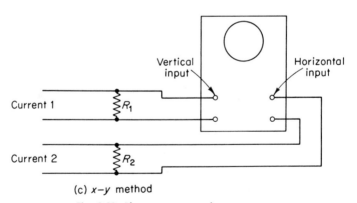

(c) *x–y* method

**Fig. 8-32.** Phase measurement between currents.

anticipated current is 10 A, the minimum wattage of the resistor should be $10^2 \times 2 = 200$ W.

### 8-15.4. Phase Measurement Between Two Currents

To measure the phase difference between two currents, pass each current through a separate fixed resistor, converting the two currents into voltages. The phase of the resulting voltages can then be compared. Two resistors capable of the necessary wattage dissipation are the only other components required for the procedure. Either the $X$-$Y$ or the dual-trace method can be used for the actual phase comparison.

Figure 8-32 shows the test connections required for converting the currents into voltages and applying both voltages to the oscilloscope.

In Fig. 8-32a, the currents are applied across corresponding resistors $R_1$ and $R_2$, with the resulting voltages applied to the two corresponding vertical inputs.

In Fig. 8-32b, the currents are applied across corresponding resistors $R_1$ and $R_2$, with the resulting voltages applied to the two corresponding electronic switch inputs.

In Fig. 8-32c, the currents are applied across corresponding resistors $R_1$ and $R_2$, with the resulting voltages applied to the vertical and horizontal inputs.

Once the test connections have been made, the phase difference can be determined by the procedures of Section 8-15.1 (dual-trace) or Section 8-15.2 ($X$-$Y$ method), whichever applies.

### NOTE

The actual resistance value of $R_1$ is not critical, but it should be low in comparison to the load. Usually 1 to 10 ohms is adequate to develop sufficient voltage for measurement. The wattage of each resistor must be at least double the square of the maximum current (in amperes). For example, if the maximum anticipated current is 3 A, the minimum wattage of the resistor should be $3^2 \times 2 = 18$ W.

# CHAPTER 9

# Solid State Measurements

## 9-1. Basic Diode Tests

Three basic tests are required for power rectifier diodes and small signal diodes. First, any diode must have the ability to pass current in one direction (forward current) and prevent or limit current flow (reverse current) in the opposite direction. Second, for a given reverse voltage, the reverse current should not exceed a given value. Third, for a given forward current, the voltage drop across the diode should not exceed a given value.

All of these tests can be made with a meter. If the diode is to be used in pulse or digital work, the switching time must also be tested. This requires an oscilloscope and pulse generator.

In addition to the basic tests, a Zener diode must also be tested for the correct Zener voltage point. Likewise, a tunnel diode must be tested for its *negative resistance* characteristics.

## 9-2. Diode Continuity Tests

The elementary purpose of a diode (both power rectifier and small signal) is to prevent current flow in one direction while passing current in the opposite direction. The simplest test of a diode is to measure current flow in the forward direction with a given voltage, then reverse the voltage and measure current flow, if any. If the diode will prevent current flow in the reverse direction but will pass current in the forward direction, the diode will meet most *basic* circuit requirements. If there is no excessive *leakage current* flow in the reverse direction, it is still possible that the diode will operate properly in noncritical circuits.

A simple resistance measurement or continuity check can often be used to test a diode's ability to pass current in one direction only. A simple ohmmeter can be used to measure the forward and reverse resistance of a diode. Figure 9-1 shows the basic circuit.

A good diode will show high resistance in the reverse direction and low resistance in the forward direction. If the resistance is high in both directions, the diode is probably open. A low resistance in both directions usually indicates a shorted diode.

It is possible for a defective diode to show a difference in forward and reverse resistance. The important factor in making a diode resistance test is the *ratio* of forward-to-reverse resistance (often known as *front-to-back* ratio). The actual ratio will depend upon the type of diode. However, as a rule of thumb, a small signal diode will have a ratio of several hundred to one, while a power rectifier can operate satisfactorily with a ratio of 10-to-1.

Diodes used in power circuits are usually not required to operate at high frequencies. Such diodes may be tested effectively with dc or low-frequency ac. Diodes used in other circuits, even audio equipment, must be capable of operation at higher frequencies and should be so tested.

There are many commercial diode testers available for use in the shop or laboratory. Most simple diode testers operate on the *continuity test* principle. This is similar to testing a diode by measuring resistance, except that actual resistance value is of no concern. Instead, arbitrary circuit values are used, and the diode's condition is read out on a "good-bad" meter.

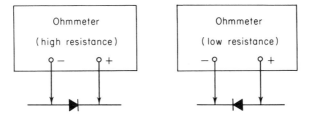

**Fig. 9-1.** Basic ohmmeter test of diodes for front-to-back ratio.

## 9-3. Diode Reverse Leakage and Forward
## Voltage Drop Tests

*Reverse leakage* is the current flow through a diode when a reverse voltage (anode negative with respect to cathode) is applied. The basic circuit for measurement of reverse leakage is shown in Fig. 9-2. Similar circuits are incorporated in some advanced diode testers or can be duplicated with the simple test equipment shown.

As shown in Fig. 9-2, the diode under test is connected to a variable d-c source in the reverse-bias condition (anode negative). The variable source is adjusted until the desired voltage is applied to the diode as indicated by the voltmeter. Then the current (if any) through the diode is measured by the current meter. This is the reverse (or leakage) current. Usually, excessive leakage current is undesired, but the limits should be determined by referring to the manufacturer's data sheet.

*Forward voltage drop* is the voltage that appears across the diode when a given forward current is being passed. The basic circuit for measurement of forward voltage is shown in Fig. 9-3. Similar circuits are incorporated in some advanced diode testers or can be duplicated with the simple test equipment shown.

As shown in Fig. 9-3, the diode under test is connected to a variable d-c source in the forward-bias condition (anode positive; cathode negative). The variable source is adjusted until the desired amount of current is passing through the diode as indicated by the current meter. Then the

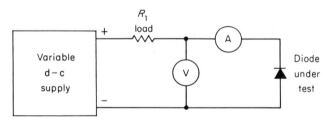

**Fig. 9-2.** D-C reverse leakage tests for diodes.

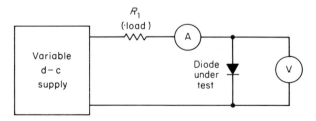

**Fig. 9-3.** D-C forward voltage drop test for diodes.

voltage drop across the diode is measured by the voltmeter. This is the forward voltage drop. Usually, a large forward voltage drop is not desired. The maximum limits should be determined by reference to the manufacturer's data sheet.

Typically, the forward voltage drop for a germanium diode will be approximately 0.2 V, while a silicon diode will have a forward drop of approximately 0.5 V.

## 9-4. Dynamic Diode Tests

The circuits and methods discussed in previous sections of this chapter provide a *static* test of diodes, meaning that the diode is subjected to constant dc when the leakage and voltage drop are measured. Diodes rarely operate this way in circuits. Instead, diodes are operated with ac, which tends to heat the diode junctions and change the characteristics. It is more realistic to test a diode under *dynamic* conditions.

### 9-4.1. Dynamic Tests for Power Rectifier Diodes

Power rectifier diodes can be subjected to a dynamic test by using a d-c oscilloscope to display and measure the current and voltage characteristics. To do so, both the vertical and horizontal channels of the oscilloscope must be *voltage calibrated*. Usually, the horizontal channel of most oscilloscopes is time calibrated. However, the horizontal channel can be voltage calibrated by using the same procedures as for the vertical channel. (Such procedures are described in the oscilloscope instruction manual.) Also, the horizontal and vertical channels must be identical or nearly identical to eliminate any phase difference.

As shown in Fig. 9-4, the power rectifier diode is tested by applying a controlled a-c voltage across the anode and cathode through resistor $R_1$. The a-c voltage (set to the maximum rated peak inverse voltage, or PIV, of the diode) alternately biases the anode positive and negative, causing both forward and reverse current to flow through $R_1$. The voltage drop across $R_1$ is applied to the vertical channel and causes the screen spot to move up and down. Vertical deflection is proportional to current. The vertical scale divisions can be converted directly to current when $R_1$ is made 1 ohm. For example, a 3-V vertical deflection indicates a 3-A current. If $R_1$ is 1000 ohms, the readout will be in milliamperes.

The same voltage applied across the diode is applied to the horizontal channel (which has been voltage calibrated), and causes the spot to move right or left. Horizontal deflection is proportional to voltage.

The combination of the horizontal (voltage) deflection and vertical

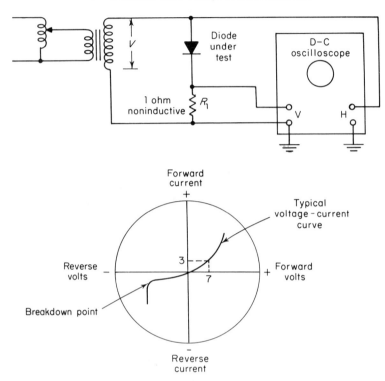

$V$ = maximum rated voltage of diode under test

**Fig. 9-4.** Testing power rectifier diode voltage-current characteristics.

(current) deflection causes the spot to trace out the complete current and voltage characteristics.

The procedure is as follows.

1. Connect the equipment as shown in Fig. 9-4.

2. Place the oscilloscope in operation. Voltage-calibrate both the vertical and horizontal channels as necessary. The spot should be at the vertical and horizontal center with no signal applied to either channel.

3. Switch off the internal recurrent sweep. Set sweep selector and sync selector to external. Leave the horizontal and vertical gain controls set at the (voltage) calibrate position.

4. Adjust the variac so that the voltage applied across the power rectifier under test is the maximum rated value.

5. Check the oscilloscope pattern against the typical curves of Fig. 9-4 and/or against the diode specifications. The curve of Fig. 9-4 is a typical response pattern. That is, the forward current increases as forward voltage

increases. Reverse current increases only slightly as reverse voltage is applied, unless the breakdown or "avalanche" point is reached. In conventional (nonZener) diodes, it is desirable if not mandatory to operate considerably below the breakdown point.

6. Compare the current and voltage values against the values specified in the diode data sheet. For example, assume that a current of 3 A should flow with 7 V applied. This can be checked by measuring along the horizontal scale to the 7-V point, then measuring from that point up (or down) to the trace. The 7-V (horizontal) point should intersect the trace at the 3-A (vertical) point.

### 9-4.2. Dynamic Tests for Small Signal Diodes

The procedures for checking the current-voltage characteristics of a small signal diode are the same as for power-rectifier diodes. However, there is one major difference. In a small signal diode, the ratio of forward voltage to reverse voltage is usually quite large. A test forward voltage of the same amplitude as the rated reverse voltage would probably damage the diode. On the other hand if the test voltage was lowered for both forward and reverse directions it would not provide a realistic value in the reverse direction.

Under ideal conditions, a small signal diode should be tested with a low-value forward voltage and a high-value reverse voltage. This can be accomplished using a circuit shown in Fig. 9-5. It will be seen that the circuit of Fig. 9-5 is essentially the same as that of Fig. 9-4 (for power rectifier diodes), except that diodes $CR_1$ and $CR_2$ are included to conduct on alternate half cycles of the voltage across transformer $T_1$. Rectifiers $CR_1$ and $CR_2$ are chosen for a linear amount of conduction near zero.

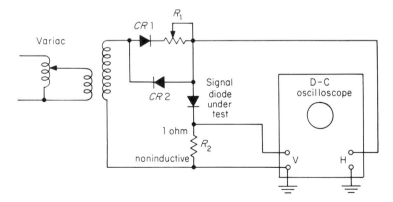

**Fig. 9-5.** Testing signal diode voltage-current characteristics.

The variac is adjusted for maximum rated reverse voltage across the diode under test as applied through $CR_2$ when the upper secondary terminal of $T_1$ goes negative. This applies the full reverse voltage.

Resistor $R_1$ is adjusted for maximum rated forward voltage across the diode as applied through $CR_1$ when the upper secondary terminal of $T_1$ goes positive. This applies a forward voltage limited by $R_1$.

With resistor $R_1$ properly adjusted, perform the current-voltage check as described for power rectifier diodes (Section 9-4.1).

## 9-5. Diode Switching Tests

Diodes to be used in pulse or digital work must be tested for switching characteristics. The most important characteristic is *recovery time*. When a reverse-bias pulse is applied to a diode, there is a measurable time delay before the reverse current reaches its steady-state value. This delay period is listed as the recovery time (or some similar term) on the diode data sheet.

The duration of recovery time sets the minimum width for pulses with which the diode can be used. For example, if a 5-$\mu$s reverse voltage pulse is applied to a diode with a 10-$\mu$s recovery time, the pulse will be distorted.

An oscilloscope having wide frequency response and good transient characteristics can be used to check the high speed switch and recovery time of diodes. The oscilloscope vertical channel must be voltage calibrated in the normal manner, while the horizontal channel should be time calibrated (rather than sweep frequency calibrated). Most laboratory oscilloscopes are time calibrated, while shop oscilloscopes are frequency calibrated.

As shown in Fig. 9-6, the diode is tested by applying a forward-biased current from the supply, adjusted by $R_2$ and measured by $M_1$. A negative square wave is developed across $R_3$. This square wave switches the diode voltage rapidly to a high negative value (reverse voltage). However, the diode does not cut off immediately. Instead, a steep transient is developed by the high momentary current flow. The reverse current falls to its steady-state value when the carriers are removed from the junction. This produces the waveform shown in Fig. 9-6.

Both forward and reverse currents are passed through resistor $R_3$. The voltage drop across $R_3$ is applied through emitter follower $Q_1$ to the oscilloscope vertical channel. The coaxial cable provides some delay so that the complete waveform will be displayed. $CR_1$ functions as a clamping diode to keep the $R_4$ voltage at a level safe for the oscilloscope.

The time interval between the negative peak and the point at which the reverse current has reached its low steady-state value is the diode recovery time.

The procedure is as follows.

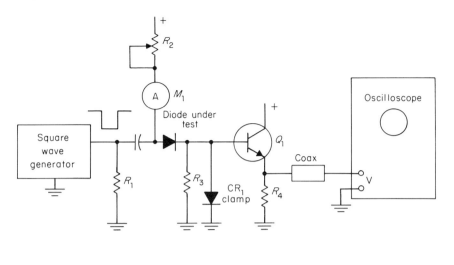

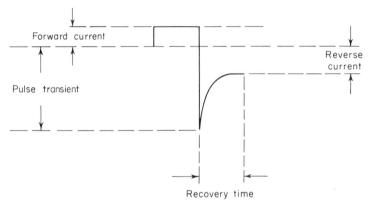

**Fig. 9-6.** Switching (recovery time) tests for diodes.

1. Connect the equipment as shown in Fig. 9-6.

2. Place the oscilloscope in operation as described in the instruction manual.

3. Switch on the internal recurrent sweep. Set sweep selector and sync selector to internal.

4. Set the square wave generator to a repetition rate of 100 kHz or as specified in the diode manufacturer's data sheet.

5. Set $R_1$ for the specified forward test current as measured on $M_1$.

6. Increase the square wave generator output level (amplitude) until a pattern appears.

7. If necessary, readjust the sweep and sync controls until a single sweep is shown.

8. Measure the recovery time along the horizontal (time-calibrated) axis.

## 9-6. Zener Diode Tests

The test of a Zener diode is similar to that of a power rectifier or small signal diode. The forward voltage drop test for a Zener is identical to that of a conventional diode, as described in Section 9-3. A reverse leakage test is usually not required, since a Zener will go into the avalanche condition when sufficient reverse voltage is applied. In place of a reverse leakage test, a Zener diode should be tested to determine the point at which avalanche occurs (establishing the Zener voltage across the diode). This can be done using a static test circuit or with a dynamic (oscilloscope) test circuit.

It is also common practice to test a Zener diode for its impedance, since the regulating ability of a Zener is related directly to the diode's impedance. A Zener diode is similar to a capacitor in this respect; as the reactance decreases so does the change in voltage across the terminals.

### 9-6.1. Static Test for Zener Diodes

The basic circuit for measurement of Zener voltage is shown in Fig. 9-7. As shown, the diode under test is connected to a variable d-c source in the reverse-bias condition (anode negative). (This is the configuration in which the Zener is normally used.) The variable source is adjusted until the Zener voltage is reached and a large current is indicated through the current meter. Zener voltage can then be measured on the voltmeter.

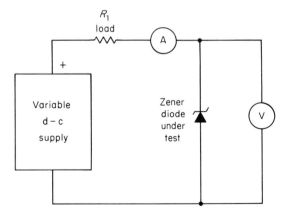

**Fig. 9-7. Basic Zener voltage test circuit.**

### 9-6.2. Dynamic Test for Zener Diodes

The procedures and circuit for dynamic test of Zener diodes are similar to those for dynamic test of conventional diodes. As shown in Fig. 9-8, the Zener diode is tested by applying a controlled a-c voltage across the anode and cathode through resistors $R_1$ and $R_2$. The a-c voltage (set to some value above the Zener voltage) alternately biases the anode positive and negative, causing both forward and reverse current to flow through $R_1$ and $R_2$.

The voltage drop across $R_2$ is applied to the vertical channel and causes the screen spot to move up and down. Vertical deflection is proportional to current. The vertical scale divisions can be converted directly to current when $R_1$ is made 1 ohm. For example, a 3-V vertical deflection indicates a 3-A current.

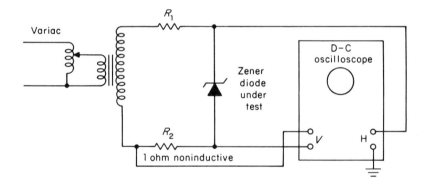

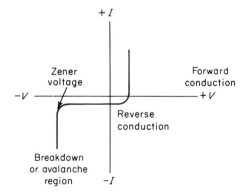

Fig. 9-8. Testing Zener diode voltage-current characteristics.

The same voltage applied across the diode (taken from the junction of $R_1$ and the diode under test) is applied to the horizontal channel (which has been voltage calibrated) and causes the spot to move right or left. Horizontal deflection is proportional to voltage. The combination of the horizontal (voltage) deflection and vertical (current) deflection causes the spot to trace out the complete current and voltage characteristics. The procedure is as follows.

1. Connect the equipment as shown in Fig. 9-8.
2. Place the oscilloscope in operation as described in the instruction manual. Voltage-calibrate both the vertical and horizontal channels as necessary. The spot should be at the vertical and horizontal center with no signal applied to either channel.
3. Switch off the internal recurrent sweep. Set sweep selector and sync selector to external. Leave the horizontal and vertical gain controls set at the (voltage) calibrate position.
4. Adjust the variac so that the voltage applied across the Zener diode and the resistors $R_1$ and $R_2$ in series is greater than the rated Zener voltage.
5. Check the oscilloscope pattern against the typical curves of Fig. 9-8 and/or against the diode specifications. The curve of Fig. 9-8 is a typical response pattern. That is, the forward current increases as forward voltage increases. Reverse (or leakage) current increases only slightly as reverse voltage is applied until the breakdown voltage is reached. Then an avalanche of current occurs.
6. Compare the current and voltage values against the values specified in the Zener diode data sheet. For example, assume that avalanche current should occur when the reverse voltage reaches 7.5 V. This can be checked by measuring along the horizontal scale up to the 7.5-V point and noting that the current increases rapidly.

### 9-6.3. Impedance Test for Zener Diodes

As discussed, the regulating ability of a Zener diode is directly related to the diode's impedance. Likewise, Zener impedance varies with junction current and diode size. Therefore, to properly test a Zener diode for impedance, measurements must be made with a specific set of conditions. This can be accomplished using the circuit of Fig. 9-9.

With such a circuit, the diode direct current is set to approximately 20% of the Zener maximum current by adjustment of $R_1$. The Zener direct current is indicated by meter $M_1$. Alternating current is also applied to the Zener, and is adjusted by $R_2$ to approximately 10% of the maximum current rating of the diode. The Zener alternating current is indicated by meter $M_3$.

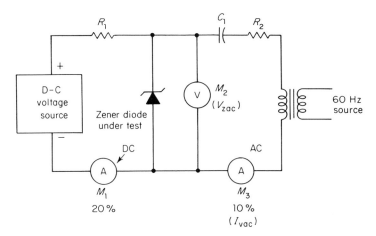

**Fig. 9-9.** Testing Zener diode impedance.

When test conditions have been met, the a-c voltage developed across the Zener junction can be read on meter $M_2$. When Zener a-c voltage ($V_{z(ac)}$) and Zener alternating current ($I_{z(ac)}$) are known, the impedance ($Z_z$) may be calculated using the equation.

$$Z_z = \frac{V_{z(ac)}}{I_{z(ac)}}$$

## 9-7. Tunnel Diode Tests

A tunnel diode must be tested for its *negative resistance* characteristics. The most effective test of a tunnel diode is to display the entire forward voltage and current characteristics on an oscilloscope. Thus, the valley and peak voltages as well as the valley and peak currents can be measured simultaneously.

It is also possible to make negative resistance tests, as well as switching tests, of tunnel diodes using meters.

### 9-7.1. Switching Test for Tunnel Diodes (Meter Method)

The basic switching test circuit for tunnel diodes is shown in Fig. 9-10. The tunnel diode under test is connected to a variable d-c source. Initially, the power source is set to zero, and is gradually increased. As the voltage is increased, there will be some voltage indication across the tunnel diode. When the critical voltage is reached, the voltage indication will "jump"

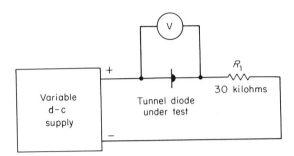

**Fig. 9-10.** Switching test for tunnel diodes (meter method).

or suddenly increase. This indicates that the diode has "switched" and is operating normally. Usually, the voltage indication will be in the order of 0.25 to 1.0 V. The power source is then decreased gradually. With a normal tunnel diode, the voltage indication will gradually decrease until a critical voltage is reached. Then the voltage indication will again "jump" and suddenly decrease.

### 9-7.2. Negative Resistance Test for Tunnel Diodes (Meter Method)

Although the negative resistance characteristics of a tunnel diode are best tested with an oscilloscope, it is possible to obtain fairly accurate results using meters connected as shown in Fig. 9-11.

Note that the diode under test is connected in the reverse-bias condition (anode negative). Therefore, any current indication on the ammeter is reverse current.

Initially, the power source is set to zero and is gradually increased until the voltage reading starts to drop (indicating that reverse current is

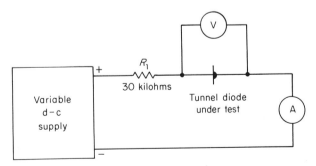

**Fig. 9-11.** Negative resistance test for tunnel diodes (meter method).

flowing and the diode is in its negative resistance region). The negative resistance region should not be confused with leakage. True negative resistance is indicated when further increases in supply voltage cause an increase in current reading but a decrease in voltage across the diode. The amount of negative resistance can be calculated using the equation

$$\text{negative resistance (in ohms)} = \frac{\text{decrease (in volts) across diode}}{\text{increase (in amperes) through diode}}$$

It is not recommended that a conventional diode be subjected to a negative resistance test unless there is a special need for the test. Also, do not operate a conventional diode in its negative resistance region any longer than is necessary. Considerable heat is generated, and the diode may be damaged.

### 9-7.3. Negative Resistance Test for Tunnel Diodes (Oscilloscope Method)

A d-c oscilloscope is required for test of a tunnel diode. Both the vertical and horizontal channels must be voltage calibrated. Also, the horizontal and vertical channels must be identical, or nearly identical, to eliminate any phase difference.

As shown in Fig. 9-12, the tunnel diode is tested by applying a controlled d-c voltage across the diode through resistor $R_3$. This d-c voltage is developed by rectifier $CR_1$ and is controlled by the variac. Current through the tunnel diode also flows through $R_3$. The voltage drop across $R_3$ is applied to the vertical channel and causes the spot to move up and down. Therefore, vertical deflection is proportional to current. Vertical scale divisions can be converted directly to current when $R_3$ is made 100 ohms. For example, a 3-V vertical deflection indicates 30 mA.

The same voltage applied across the tunnel diode is applied to the horizontal channel (which has been voltage calibrated) and causes the spot to move from left to right. (For a tunnel diode test, the horizontal and vertical zero-reference point should be at the *lower left* of the screen rather than in the center.) The horizontal deflection is proportional to voltage. The combination of the horizontal (voltage) deflection and vertical (current) deflection causes the spot to trace out the complete negative resistance characteristic.

1. Connect the equipment as shown in Fig. 9-12.

2. Place the oscilloscope in operation as described in the instruction manual. Voltage-calibrate both the vertical and horizontal channels as necessary. The spot should be at the lower left-hand side of center with no signal applied to either channel.

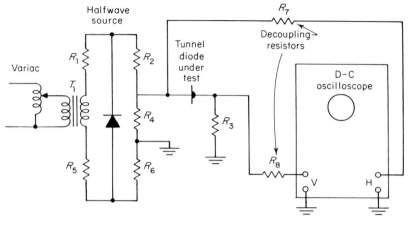

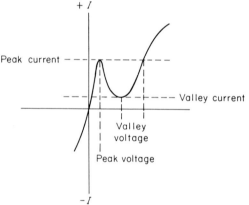

**Fig. 9-12.** Negative resistance test for tunnel diodes (oscilloscope method).

3. Switch off the internal recurrent sweep. Set sweep selector and sync selector to external. Leave the horizontal and vertical gain controls set at the (voltage) calibration position.

4. Adjust the variac so that the voltage applied across the tunnel diode under test is the maximum rated forward voltage. This can be read across the voltage-calibrated horizontal axis.

5. Check the oscilloscope pattern against the curve of Fig. 9-12 or against the tunnel diode manufacturer's data.

6. The following equation can be used to obtain a *rough approximation* of negative resistance in tunnel diodes.

$$\text{negative resistance} = \frac{E_v - E_p}{2(I_p - I_v)}$$

where $E_v$ is valley voltage,
     $E_p$ is peak voltage,
     $I_v$ is valley current, and
     $I_p$ is peak current

## 9-8. Basic Transistor Testing

There are four basic tests required for transistors in practical applications: gain, leakage, breakdown, and switching time. All of these tests are best made with commercial transistor testers and/or oscilloscopes. However, it is possible to test a transistor with an ohmmeter. These simple ohmmeter tests will show if the transistor has leakage and if the transistor shows some gain.

In the final analysis, the only true test of a transistor is in the circuit with which the transistor is to be used. However, except in special circumstances, a transistor will operate properly "in-circuit" provided that the transistor shows the proper gain, that it does not break down under the maximum operating voltages, and that the leakage is within tolerance and, in the case of pulse circuits, the switching characteristics (such as delay time, storage time, etc.) are within tolerance.

There are two exceptions to this rule. Transistor characteristics will change with variations in operating *frequency* and *temperature*. For example, a transistor may be tested at 1 MHz and show more than enough gain to meet circuit requirements. However, at 10 MHz, the gain of the same transistor may be zero. This can be due to a number of factors. Any transistor will have some capacitance at the input and the output. As frequency increases, the capacitive reactance will change until at some frequency the transistor will become unsuitable for the circuit. In the case of temperature, the current flow in any junction will increase with increases in temperature. A transistor may be tested for leakage at a normal ambient temperature and show a leakage well within tolerance. When the same transistor is used "in-circuit" the temperature will increase, increasing the leakage to an unsuitable level.

It is usually not practical to test transistors over the entire range of operating frequencies and temperatures with which the transistor will be used. Instead the transistor should be tested under the conditions specified in the data sheet. Then, by using equations and graphs, the transistor characteristics can be predicted at other frequencies and temperatures. (Such equations and graphs, together with an explanation of their use, and a discussion concerning interpretation of transistor test results are given in the author's *Practical Semiconductor Data Book for Electronic Engineers and Technicians* (Englewood Cliffs, N.J., Prentice-Hall, Inc., 1969).)

## 9-9. Transistor Tests (Ohmmeter Method)

The following tests can be performed with an ohmmeter and do not require an oscilloscope or commercial transistor testers.

### 9-9.1. Transistor Leakage Tests (Ohmmeter Method)

For meter test purposes, transistors can be considered as two diodes connected back-to-back. Therefore, each diode should show low forward resistance and high reverse resistance. These resistances can be measured with an ohmmeter as shown in Fig. 9-13. The same ohmmeter range should be used for each pair of measurements (base-to-emitter, base-to-collector, and collector-to-emitter).

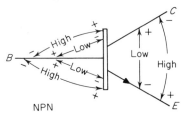

NPN

On low-power transistors, there may be a few ohms indicated from collector-to-emitter. Avoid using the $R \times 1$ range or an ohmmeter with a high internal battery voltage. Either of these conditions can damage a low-power transistor.

If both forward and reverse readings are very high, the transistor is open. Likewise, if any of the readings show a short or very low resistance, the transistor is shorted. Also, if the forward and reverse readings are the same (or nearly equal), the transistor is defective.

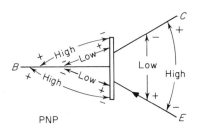

PNP

**Fig. 9-13.** Transistor leakage tests (ohmmeter method).

A typical forward resistance is 300 to 700 ohms. Typical reverse resistances are 10 to 60 kilohms. Actual resistance values will depend upon ohmmeter range and battery voltage. Therefore, the *ratio of forward-to-reverse* resistance is the best indicator. Almost any transistor will show a ratio of at least 30-to-1. Many transistors show ratios of 100-to-1 or greater.

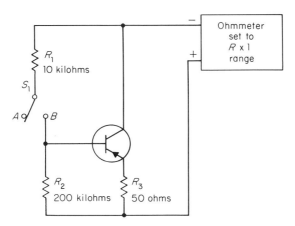

Fig. 9-14. Transistor gain test (ohmmeter method).

### 9-9.2. Transistor Gain Tests (Ohmmeter Method)

Normally, there will be little or no current flow between emitter and collector until the base-emitter junction is forward biased. Therefore, a basic gain test of a transistor can be made using an ohmmeter. The test circuit is shown in Fig. 9-14.

In this test, the $R \times 1$ range should be used. Any internal battery voltage can be used provided that it does not exceed the maximum collector-emitter breakdown voltage.

In position $A$ of switch $S_1$, there is no voltage applied to the base, and the base-emitter junction is not forward biased. Therefore, the ohmmeter should read a high resistance. When switch $S_1$ is set to $B$, the base-emitter circuit is forward biased (by the voltage across $R_1$ and $R_2$), and current flows in the emitter-collector circuit. This is indicated by a lower resistance reading on the ohmmeter. A 10-to-1 resistance ratio is typical for an audio frequency transistor.

## 9-10. Transistor Leakage Tests

In theory, there should be no collector-base current flow, since the collector-base junction of a transistor is reverse biased in a normal circuit. In practical applications, however, there will be some collector-base current flow, or *collector leakage current*. This is designated as $I_{CBO}$ or $I_{CO}$ on most data sheets. Collector leakage can also be termed "collector cutoff

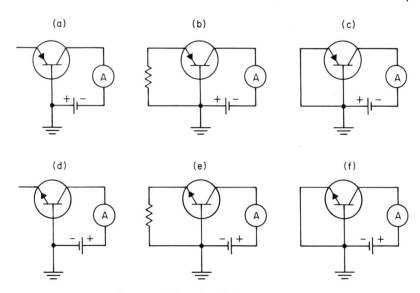

**Fig. 9-15.** Collector-base leakage test circuits.

current," where a nominal and/or maximum current is specified for a given collector-base voltage and ambient temperature.

Figure 9-15 shows the basic circuits for collector-base leakage tests. Although any of the circuits could be used, those of Figs. 9-15a and 9-15b (emitter open) are the most popular and the most accurate.

The procedure is the same for all of the circuits shown in Fig. 9-15. The voltage source is adjusted to a given value (thus providing a given reverse bias), and the current (if any) is read on the meter. This current must be *below a given maximum for a given reverse bias.*

Temperature is often a critical factor in leakage measurements. Always make leakage measurements at the specified data sheet temperature.

Emitter-base current leakage ( $I_{EO}$ or $I_{EBO}$) is sometimes specified. In that event the circuits and procedures of Fig. 9-15 can be used, except that the collector and emitter connections are interchanged (emitter-base reverse biased; collector open).

## CAUTION

Transistors should not be tested for any characteristic unless all of the characteristics are known. Never test a transistor with voltages or currents higher than the rated values. Pay particular attention to *maximum current rating.* For example, if a maximum 45 V is specified for the collector, it could be assumed that a 9-V battery would be safe for all measurements. However, assuming that the emitter-to-collector resistance was 90 ohms and a maximum rated collector current was 25 mA, this maximum would be exceeded. To place 9 V directly across emitter and collector would produce 100 mA (four times the 25-mA limit).

## 9-11. Transistor Breakdown Tests

Collector-base breakdown voltage is the most often specified value. In this test, collector and base are reverse biased with emitter open. The voltage source is adjusted to a *given value of leakage current;* then the voltage is compared with the specified minimum collector breakdown voltage. If the specified current will flow with a lower voltage, the collector-base junction is breaking down.

Sometimes, collector-emitter breakdown voltage is specified (collector-emitter reverse biased; base open). The voltage source is then adjusted for a given value of leakage current through *both* the emitter-base and collector-base junctions. The collector-emitter breakdown test determines the condition of both junctions simultaneously.

Breakdown voltage is designated as $BV_{CBO}$ (collector-base, emitter open), $BV_{CES}$ (emitter shorted to base), or $BV_{CEO}$ (collector-to-emitter, base open) on most data sheets. Breakdown is normally measured with the emitter shorted to the base, connected to the base through a resistor, or with the emitter and base reverse biased.

Figure 9-16 shows the basic circuits for breakdown tests. The circuits shown are for PNP transistors. The same circuits can be used for NPN transistors when the voltage source polarity is reversed.

In all cases, the voltage source is adjusted for a given leakage current flow. Then the voltage is compared with a minimum specified voltage.

## 9-12. Transistor Gain Tests

The dynamic gain of a transistor is determined by the amount of change in output for a given change in input. Usually, transistors are tested for *current gain*. The change in output current for a given change in input current is measured without changing the output voltage.

Common-base current gain is known as alpha ($\alpha$). Common-emitter current gain is known as beta ($\beta$). Present-day data sheets also use several other terms to specify gain. The term "forward current transfer ratio" and the letters "$h_{fe}$" are the most popular means of indicating current gain, though some manufacturers use "collector-to-base current gain."

The "$h$" in the letters "$h_{fe}$" refers to the hybrid (mixed) arrangement of transistor equivalent model circuits, where the transistor and test or operating circuit are considered as a "black box" with an input and an output rather than individual components.

Lowercase letters $h_{fe}$ (sometimes $H_{fe}$) indicate that the current gain

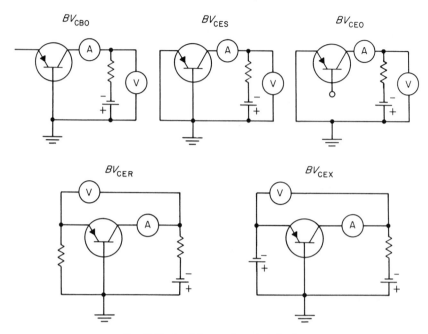

**Fig. 9-16.** Breakdown voltage test circuits.

is measured by noting the change in collector alternating current for a given change in base alternating current. This is also known as "a-c beta" or "dynamic beta."

Capital letters $H_{FE}$ indicate that current gain is measured by noting the collector direct current for a given base direct current. This is also known as "d-c beta."

Direct-current gain measurements apply under a wider range of conditions and are easier to make. Alternating-current gain measurements require more elaborate test circuits, and the test results will vary with the frequency of the ac used for test. However, a-c measurements are more realistic since transistors are usually used with a-c signals.

### 9-12.1. Basic Transistor Gain Tests

Figure 9-17 shows the basic circuits for alpha measurements of PNP and NPN transistors. Under static conditions, both the emitter current $I_E$ and collector current $I_C$ are measured. Then the emitter current $I_E$ is changed a given amount by varying the resistance of $R_1$ or by changing the emitter-base source voltage. The collector-to-base voltage must remain the same.

The *difference* in collector current $I_C$ is noted, and the value of alpha

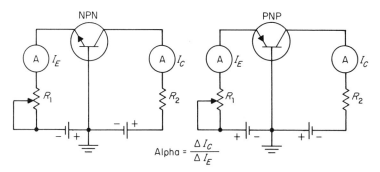

**Fig. 9-17.** Basic alpha test circuits.

is calculated using the equation shown in Fig. 9-17. For example, assume that the emitter current $I_E$ is changed 4 mA and that this results in a change of 1 mA in collector current $I_C$. This would mean a current gain of 0.25.

Figure 9-18 shows the basic circuits for beta measurements of PNP and NPN transistors. Under static conditions, both the base current $I_B$ and the collector current $I_C$ are measured. Then, without changing the collector voltage, the base current $I_B$ is changed by a given amount, and the difference in collector current $I_C$ is noted.

For example, assume that when the circuit is first connected $I_B$ is 7 mA and $I_C$ is 43 mA. When $I_B$ is increased to 10 mA (a 3-mA increase), $I_C$ increases to 70 mA (a 27-mA increase). This represents a 27-mA increase in $I_C$ for a 3-mA increase in $I_B$ or a current gain of 9 mA.

### NOTE

Certain precautions should be observed if transistors are to be tested using *noncommercial* test circuits. The most important of these are as follows.

1. The collector and emitter (or base) load resistances (represented by $R_1$ and $R_2$ in Figs. 9-17 and 9-18) should be of such value that the maximum current limitations of the transistor are not exceeded (as discussed in Section 9-10). In the case of power transistors, the wattage rating of the load resistance should be large enough to dissipate the heat.

2. A large collector-leakage current must be accounted for in test conditions.

3. The effect of meters used in the test circuits must also be accounted for in test conditions.

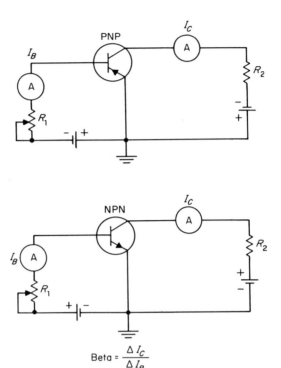

$$Beta = \frac{\Delta I_C}{\Delta I_B}$$

**Fig. 9-18.** Basic beta test circuits.

### 9-12.2. Commercial Transistor Testers

There are several types of circuits used in commercial transistor testers to measure a-c or dynamic gain. Some of the commercial transistor testers use the same basic circuits shown in Figs. 9-17 and 9-18, except that an a-c signal is introduced into the input and the gain is measured at the output. Usually, these testers provide a 60-Hz or 1000-Hz signal for injection into the input. When it is desirable to test transistors at higher frequencies, some tester circuits permit an external high-frequency signal to be injected.

One of the most common methods used in transistor testers for a-c gain measurement is the *feedback* circuit. In such a circuit, the transistor under test is inserted into an audio oscillator configuration. The amount of feedback is adjusted by means of a calibrated control until the circuit barely starts to oscillate. Oscillation is indicated by a tone on the loudspeaker. The transistor current gain at the oscillation starting point can be read directly from the dial calibration.

### 9-12.3. Laboratory Transistor A-C Gain Tests

The most practical means of measuring transistor gain in the laboratory is to display the transistor characteristics as curve traces on an oscilloscope. Since the oscilloscope screen can be calibrated in voltage and current, the transistor characteristics can be read off the screen directly. If a number of curves are made with an oscilloscope, they can be compared with the curves drawn on transistor data sheets.

There are a number of oscilloscopes (or oscilloscope adapters) manufactured specifically to display transistor curves. The Tektronix Transistor Curve-tracer is a typical unit. Transistors under test are inserted into either a common-base or common-emitter test circuit. The transistor collector has a sweep voltage applied to it while a step voltage is applied to either the base or emitter (whichever is ungrounded). Voltage (for the collector) sweeps between zero and a selected value. Step voltages (for the emitter or base) start at zero and build up to a value determined by the number of steps and value of each step as selected. Each sequence of steps, from zero to the maximum attained value, in conjunction with the sweep voltage on the collector, produces one family of characteristic curves.

Signals used for vertical and horizontal deflection on the oscilloscope screen are either current or voltage values selected from various points in the transistor test circuit. Thus, a selected vertical signal can be plotted against a selected horizontal signal to trace the desired transistor characteristic curve. A laboratory curve-tracer contains circuits that permit almost any combination of current and collector, emitter, or base voltage to be displayed.

These circuits can be duplicated using a d-c oscilloscope for display. The most important set of curves are those that show *collector output versus emitter input*. Figure 9-19 illustrates the basic circuit required for such a display. For this circuit, both the vertical and horizontal channels of the oscilloscope must be voltage calibrated. Usually, the horizontal channel of an oscilloscope is calibrated with respect to time. The horizontal and vertical channels must be identical, or nearly identical, to eliminate any phase difference. The horizontal-zero reference point should be at the left (or right) of the oscilloscope screen rather than in the center.

In Fig. 9-19 the transistor is tested by applying a controlled d-c voltage to the collector. The collector voltage is developed by rectifying the transformer $T_1$ secondary voltage with diode $CR_1$ and can be adjusted to any desired value by the variac. When collector current flows on positive half cycles, the current flows through $R_1$. The voltage drop across $R_1$ is applied

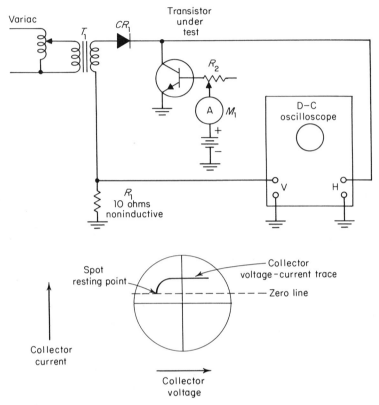

**Fig. 9-19.** Testing NPN transistors for collector current versus input current (common emitter).

to the vertical channel and causes the spot to move up and down. Therefore, vertical deflection is proportional to current. With $R_1$ at a value of 10 ohms as shown in Fig. 9-19, the indicated voltage value must be divided by ten to obtain current. For example, a 3-V vertical deflection indicates a 0.3-A (300-mA) current.

The same voltage applied to the transistor collector is applied to the horizontal channel (which has been voltage calibrated) and causes the spot to move from left to right (for the NPN transistor shown). Therefore, horizontal position is proportional to voltage. The combination of the horizontal (voltage) deflection and vertical (current) deflection causes the spot to trace out the collector-current and collector-voltage characteristics. Usually, the change in collector current for a given change in emitter-base current (beta) is the desired characteristic for most transistors. This can be displayed by setting the emitter-base current to a given value and measuring the collector-current curve with a given collector voltage. Then the emitter-base current is changed to another value, and the new collector current is displayed without changing the collector volt-age.    Collector voltage is set by the variac. Emitter-base current is set by $R_2$ and measured on $M_1$. On the commercial transistor curve-tracers, the emitter-base current is applied in steps.

The test connection diagram of Fig. 9-19 is for an NPN transistor connected in a common emitter circuit. If a PNP transistor is to be tested, the polarity of rectifier $CR_1$, Battery $B$, and meter $M_1$ must be reversed. Also, the horizontal zero-reference point should be at the right of the screen rather than at the left.

The following procedure will display a *single curve*.

1. Connect the equipment as shown in Figure 9-19.

2. Place the oscilloscope in operation as described in the instruction manual. Voltage-calibrate both the vertical and horizontal channels as necessary. The spot should be at the vertical center and at the left (for NPN) of the horizontal center with no signal applied to either channel.

3. Switch off the internal recurrent sweep. Set sweep selector and sync selector to external. Leave the horizontal and vertical gain controls set at the (voltage) calibrate position. Set the vertical polarity switch so that the trace will deflect up from the center line as shown in Fig. 9-19.

4. Adjust the variac so that the voltage applied to the collector is the maximum rated value. This voltage can be read on the voltage-calibrated horizontal axis.

5. Adjust resistor $R_2$ for the desired emitter-base current as indicated on meter $M_1$.

6. Check the oscilloscope pattern against the transistor specifications. Compare the current-voltage values against the specified values. For ex-ample, assume that a collector current of 300 mA should flow with 7 V applied. This can be checked by measuring along the horizontal scale to the 7-V point, then measuring from that point up the trace. The 7-V (horizontal) point should intersect the trace at the 300-mA (3 V) vertical point.

7. If desired, adjust resistor $R_2$ for another emitter-base current value as indicated on meter $M_1$. Then check the new collector current and voltage curve.

## 9-13. Transistor Switching Tests

Transistors to be used in pulse or digital applications must be tested for switching characteristics. For example, when a pulse is applied to the input of a transistor, there is a measurable time delay before the pulse starts to appear at the output. After the pulse is removed, there is addi-tional time delay before the transistor output returns to the normal level.

These "switching times" or "turn on" and "turn off" times are usually in the order of a few microseconds for high-speed pulse transistors.

The switching characteristics of transistors designed for computer or industrial work are listed on the data sheets. Each manufacturer lists his own set of specifications. However, there are four terms (rise time, fall time, delay time, and storage time) common to most data sheets for transistors used in pulse work. These switching characteristics are of particular importance when the pulse durations are short. For example, assume that the "turn on" time of a transistor is 10 $\mu$s and that a 5-$\mu$s pulse is applied to the transistor input. Under these conditions there would be no output pulse, or the pulse would be drastically distorted.

### 9-13.1. Pulse and Square Wave Definitions

The following terms are commonly used in describing transistor switching characteristics. The terms are illustrated in Fig. 9-20. The input pulse represents an ideal input waveform for comparison purposes. The other waveworms represent typical output waveforms. In order to show the relationships the terms are defined as follows.

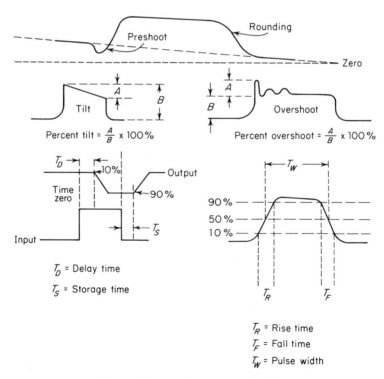

**Fig. 9-20.** Pulse and square wave definitions.

1. *Rise time $T_R$* is the time interval during which the amplitude of the output voltage changes from 10% to 90% of the rising portion of the pulse.

2. *Fall time $T_F$* is the time interval during which the amplitude of the output voltage changes from 90% to 10% of the falling portion of the waveform.

3. *Time delay $T_D$* is the time interval between the beginning of the input pulse (time zero) and the time when the rising portion of the output pulse attains an arbitrary amplitude of 10% above the baseline.

4. *Storage time $T_S$* is the time interval between the end of the input pulse (trailing edge) and the time when the falling portion of the output pulse drops to an arbitrary amplitude of 90% from the baseline.

5. *Pulse width* (or pulse duration) $T_W$ is the time duration of the pulse measured between two 50% amplitude levels of the rising and falling portions of the waveforms.

6. *Tilt* is a measure of the tilt of the full amplitude flattop portion of a pulse. The tilt measurement is usually expressed as a percentage of the amplitude of the rising portion of the pulse.

7. *Overshoot* is a measure of the overshoot generally occurring above the 100% amplitude level. This measurement is also expressed as a percentage of the pulse rise.

These definitions are for guide purposes only. When pulses are very irregular (such as excessive tilt, overshoot, etc.), the definitions may become ambiguous.

### 9-13.2. Testing Transistors for Switching Time

An oscilloscope having wide frequency response, good transient characteristics, and a dual trace can be used to check the high-speed switching characteristics of transistors used in pulse or computer work. The oscilloscope vertical channel must be voltage calibrated in the normal manner, whereas the horizontal channel should be time calibrated (rather than sweep frequency calibrated).

As shown in Fig. 9-21, the transistor is tested by applying a pulse to the base of the transistor under test. This same pulse is applied to one of the oscilloscope vertical inputs. The transistor collector output is applied to the other oscilloscope vertical input (inverted 180° by the common-emitter circuit). The two pulses are then compared as to rise time, fall time, delay time, storage time, etc. The transistor output pulse characteristics can then be compared with the transistor specifications.

1. Connect the equipment as shown in Fig. 9-21.

2. Place the oscilloscope in operation as described in the instruction

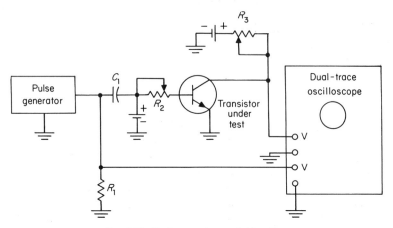

**Fig. 9-21.** Testing transistor switching time.

manual. Switch on the oscilloscope internal recurrent sweep. Set the sweep selector and sync selector to internal.

3. Set the pulse generator to produce a 2.2-V, 3-$\mu$s positive pulse. These values are taken (arbitrarily) from the data sheet of a General Electric 2N338 transistor but are typical for many computer transistors. Always use the values specified in the data sheet.

4. Adjust the collector and base supply voltages to the values specified in the transistor manufacturer's data.

5. The oscilloscope pattern should appear as shown in Fig. 9-20, with both the transistor output pulse and input pulse displayed.

6. Measure the rise time, fall time, delay time, and storage time along the horizontal (time calibrated) axis of the oscilloscope display.

### 9-13.3. Rule of Thumb for Switching Tests

Since rise-time (and fall-time) measurements are of special importance in switching tests, the relationship between oscilloscope rise time and the rise time of the transistor (or other device such as a diode) must be taken into account. Obviously, the accuracy of rise-time measurements can be no greater than the rise time of the oscilloscope. Also, if the device is tested by means of an external pulse from a pulse generator, the generator's rise time must also be taken into account.

For example, if an oscilloscope with a 20-ns rise time is used to measure the rise time of a 15-ns transistor, the measurement would be hopelessly inaccurate. If a 20-ns pulse generator and a 15-ns oscilloscope were used to measure the rise time of a device, the fastest rise time for accurate measurement would be something greater than 20 ns.

There are two basic rules of thumb that can be applied to rise-time measurements. The first method is known as the "root of the sum of the squares." It involves finding the square of all the rise times associated with the test, adding these squares together, and then finding the square root of this sum.

For example, when using the 20-ns pulse generator and the 15-ns oscilloscope the calculation would be

$$20 \times 20 = 400, \qquad 15 \times 15 = 225, \qquad 400 + 225 = 625,$$

$$\sqrt{625} = 25 \text{ ns}$$

One problem with this rule is that the coaxial cables required to interconnect the test equipment are subject to "skin effects." As frequency increases, the signals tend to travel on the outside or skin of the conductor. This decreases conductor area and increases resistance, which in turn increases cable loss. Cable loss does not add properly to enable application of the root-sum-squares method except as an approximation.

The second method states that if the equipment or signal being measured has a rise time *ten times* slower than the test equipment, the error is 1%. This is small and can be ignored for practical testing. If the equipment being measured has a rise time *three times* slower than the test equipment, the error is slightly less than 6%.

## 9-14. Testing Unijunction Transistors

As in the case of conventional junction transistors, there are a number of commercial test sets for UJTs, and it is possible to adapt commercial test sets (designed to test conventional junction transistors) for use with UJTs. The following describes circuits and procedures for testing of all major UJT characteristics.

### 9-14.1. Simplified Test for UJTs

The firing point of a UJT can be determined using a simple ammeter circuit. The test will also show the amount of emitter-base-one current flow after the UJT is fired. If a UJT will fire with the correct voltage applied and draws the rated amount of current, it can be considered satisfactory for operation in most circuits.

The test circuit is shown in Fig. 9-22. The base-two voltage is shown as $+20$ V. However, any value of base-two voltage can be used to match a particular UJT.

Initially, $R_1$ is set to 0 V (at the ground end). The setting of $R_1$ is gradually increased until the unijunction fires. The firing voltage is indicated on the voltmeter. When the UJT fires, the ammeter indication will increase suddenly. The amount of emitter-base-one current is read on

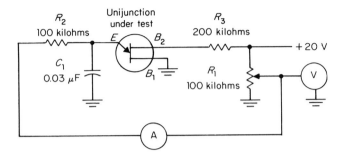

**Fig. 9-22. Unijunction transistor firing test.**

the ammeter. Usually, the firing potential is on the order of 0 to 20 V, whereas current is less than 50 μA.

### 9-14.2. Emitter and Interbase Curve Trace for UJTs

The circuit of Fig. 9-23 can be used to display either the emitter characteristic curves or the interbase characteristic curves on a standard oscilloscope. The meter indicates emitter current (with the switches set to display interbase curves), or interbase voltage (with the switches set to display emitter curves).

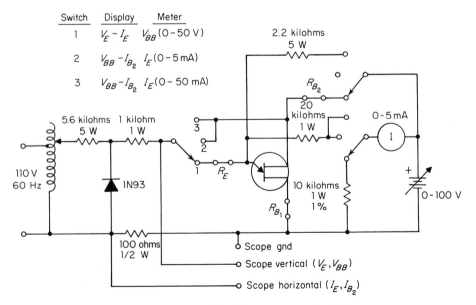

**Fig. 9-23. Emitter and interbase curve tracer for UJTs.**

It is important to set the variac and the d-c supply to zero *before* changing the switches or inserting the UJT in the circuit so as to avoid accidental burnout.

If desired, external resistors may be inserted in series with the emitter, base one, or base two to determine the UJT characteristics in a particular circuit.

### 9-14.3. Commercial Test Equipment for UJTs

A unit such as the Tektronix 575 transistor curve-tracer can also be used to display the characteristic curves of the UJT. Connections that may be used with the Tektronix 575 are shown in Fig 9-24. In the display of the emitter characteristic curves, the interbase voltage cannot exceed 12 V because of the voltage limitations of the base current step generator.

## 9-15. Testing Field Effect Transistors

As in the case of most electronic components, FETs are subjected to a variety of tests during manufacture. It is neither practical nor necessary to duplicate all of these tests in the field. FETs can be tested under static

| Test | Emitter curves | Interbase curves |
|---|---|---|
| Circuit | 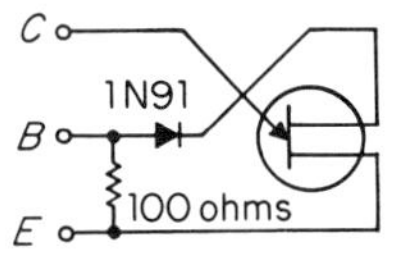 | 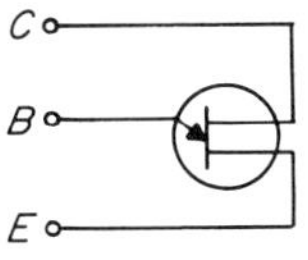 |
| Collector sweep polarity | + | + |
| Base step polarity | + | + |
| Collector peak volt range | 200 V | 20 V |
| Collector limiting resistor | 5 kilohms | 500 ohms |
| Base current steps selector | 20 $\mu$A/step | 10 $\mu$A/step |
| Number of current steps | 5 | 5 |
| Vertical current range | 2 $\mu$A/div $I_E$ | 2 $\mu$A/div $I_{B_2}$ |
| Horizontal voltage range | 1 V/div $V_E$ | 2 V/div $V_{BB}$ |

**Fig. 9-24.** Use of Tektronix 575 transistor curve tracer for displays of UJT characteristic curves.

or dynamic conditions. Static (or d-c) tests include gain (the value of output current for a given voltage input), gate leakage, and voltage breakdown.

In general, it is more practical to test an FET under dynamic conditions. Unlike the static characteristics, dynamic characteristics (a-c or signal) of FETs apply equally to all types (*A*, depletion only; *B*, depletion/enhancement; *C*, enhancement only). However, the conditions and presentations of the dynamic characteristics depend mostly on the intended *application*. Table 9-1 indicates the dynamic characteristics needed to describe an FET for various applications.

**TABLE 9-1**

Typical FET Dynamic Characteristics

| Audio | RF-IF | Switching | Chopper |
|-------|-------|-----------|---------|
| $Y_{fs}$ (1 kHz) | $Y_{fs}$ (1 kHz) | $Y_{fs}$ (1 kHz) | $Y_{fs}$ (1 kHz) |
| $C_{iss}$ | $C_{iss}$ | $C_{iss}$ | $C_{iss}$ |
| $C_{rss}$ | $C_{rss}$ | $C_{rss}$ | $C_{rss}$ |
| $Y_{os}$ (1 kHz) | $Y_{fs}$ (HF) | $C_{d(\text{sub})}$ | $C_{d(\text{sub})}$ |
| $Y_{fs}$ (HF) | Re $(Y_{is})$ (HF) | $r_{ds(\text{on})}$ | $r_{ds(\text{on})}$ |
| NF | Re $(Y_{os})$ (HF) | $t_{d1}, t_{d2}$ | |
| | NF | $t_r, t_f$ | |

### 9-15.1. Forward Transadmittance (or Transconductance) $Y_{fs}$ Test for FETs

This test is for the most important dynamic characteristic of FETs, no matter what the application. $Y_{fs}$ serves as a basic design parameter in audio and RF and is a widely accepted device figure of merit.

Because FETs have many characteristics similar to those of vacuum tubes, the symbol $g_m$ is sometimes used instead of $Y_{fs}$. This is further confused, since the $g$ notation school also used a number of subscripts. In addition to $g_m$, some data sheets show $g_{fs}$, while others go even further with $g_{21}$.

No matter what symbol is used, $Y_{fs}$ defines the relationship of *output signal current* divided by *input signal voltage,* with the drain-source voltage held constant. $Y_{fs}$ is expressed in ohms. Figure 9-25 is a typical $Y_{fs}$ test circuit for a tetrode-connected JFET.

$Y_{fs}$ is usually specified at 1 kHz, with a drain-source voltage that will produce some specified drain current. Since the drain-current gate voltage curve of an FET is nonlinear, $Y_{fs}$ will vary considerably with changes in drain current. This variation for a typical n-channel Type *A* JFET (3N126) is illustrated in Fig. 9-26.

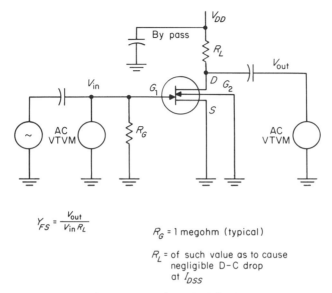

$$Y_{FS} = \frac{V_{out}}{V_{in} R_L}$$

$R_G$ = 1 megohm (typical)

$R_L$ = of such value as to cause negligible D–C drop at $I_{DSS}$

**Fig. 9-25.** Typical test circuit for $Y_{fs}$.

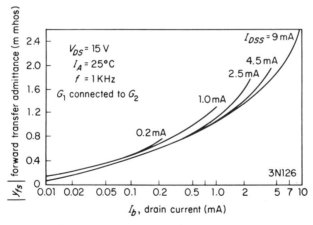

**Fig. 9-26.** Forward transfer admittance versus drain current for typical 3N126 JFET.

Three $Y_{fs}$ measurements are usually specified for tetrode-connected FETs. One of these, with two gates tied together, provides a $Y_{fs}$ value for the condition in which a signal is applied to both gates simultaneously. The other two measurements provide the $Y_{fs}$ for the two gates individually. Generally, with the two gates tied together, $Y_{fs}$ is higher, and more gain may be realized in a given circuit. However, because of the increased capacitance, the gain-bandwidth product is much lower.

For FETs used at radio frequencies, an additional value of $Y_{fs}$ should be specified at or near the *highest* frequency of operation. This value should also be measured at the same voltage conditions as those used to produce the specified drain current.

The real portion of this high-frequency $Y_{fs}$, such as $\mathrm{Re}(Y_{fs})(\mathrm{HF})$ or $g_{21}$, is usually considered as a significant figure of merit.

### 9-15.2. Output Admittance $Y_{os}$ Test for FETs

$Y_{os}$ is also represented by various $Y$ and $g$ parameters, such as $Y_{22}$, $g_{os}$, and $g_{22}$. $Y_{os}$ is even specified in terms of drain resistance or $r_d$, where $r_d = 1/Y_{os}$. This is similar to the vacuum tube characteristic of output admittance, where output admittance is equal to one divided by plate resistance.

No matter what symbol is used, $Y_{os}$ defines the relationship of *output drain current* divided by *output drain voltage,* with the input gate voltage held constant. $Y_{os}$ is expressed in ohms. Figure 9-27 is a typical $Y_{os}$ test circuit for Types $A$ and $B$ FETs, which are measured with the gates and source grounded. For Type $C$ FETs, $Y_{os}$ is measured at some specified value of drain-source voltage that will permit a large drain-current flow.

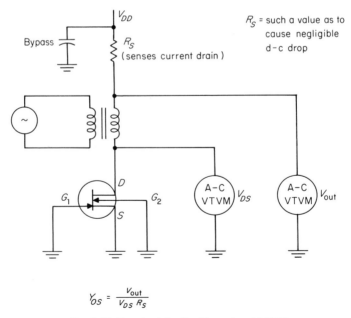

$$Y_{os} = \frac{V_{out}}{V_{DS}\,R_S}$$

**Fig. 9-27.** Test circuit for $Y_{os}$ (Types A and B FETs).

Voltages and frequencies for measuring $Y_{os}$ should be the same as for $Y_{fs}$.

### 9-15.3. Input Capacitance Tests for FETs

$C_{iss}$ (input capacitance with a common source circuit) is used on specification sheets for low-frequency FETs. $Y_{is}$ (input admittance with a common source circuit) is more popular for RF FETs. Figure 9-28 shows typical circuits for tests of $C_{iss}$ in tetrode JFETs. As with $Y_{fs}$, two measurements (one with gate 2 tied to source, and the other with common gates) are necessary for tetrode JFETs. $C_{iss}$ is of particular importance when FETs are used in switching circuits, since a large voltage swing will occur across the input capacitance.

### 9-15.4. Reverse Transfer Capacitance
### Tests for FETs

$C_{rss}$ (reverse transfer capacitance with a common source circuit) is used on specification sheets for both low- and high-frequency FETs. $Y_{rs}$ (reverse transfer admittance) is rarely used. This is because $Y_{rs}$ for an

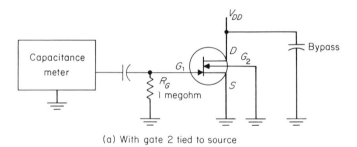

(a) With gate 2 tied to source

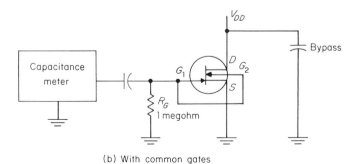

(b) With common gates

**Fig. 9-28.** Test circuit for $C_{iss}$ in tetrode JFETs.

FET remains almost completely capacitive and of relatively constant capacity over the entire usable FET frequency spectrum. Consequently, the low-frequency $C_{rss}$ is an adequate specification.

Figure 9-29 shows typical circuits for tests of $C_{rss}$ in tetrode JFETs. Again, two measurements (gate 1 individually, and common gates) are necessary for tetrode JFETs.

$C_{rss}$ is also important in switching applications. For chopper circuits, $C_{rss}$ is actually the feed-through capacitance for the chopper drive. $C_{rss}$ is also known as the Miller-effect capacitance, since the reverse capacitance

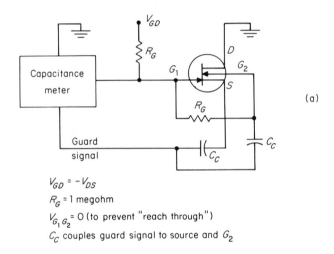

(a)

$V_{GD} = -V_{DS}$

$R_G = 1$ megohm

$V_{G_1 G_2} = 0$ (to prevent "reach through")

$C_C$ couples guard signal to source and $G_2$

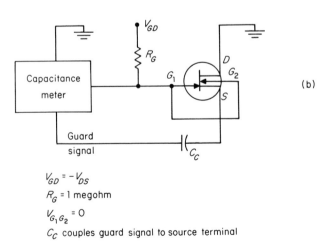

(b)

$V_{GD} = -V_{DS}$

$R_G = 1$ megohm

$V_{G_1 G_2} = 0$

$C_C$ couples guard signal to source terminal

**Fig. 9-29.** Test circuit for $C_{rss}$.

can produce a condition similar to the Miller effect in vacuum tubes (such as continually changing frequency-response curves).

### 9-15.5. Channel Resistance $r_{ds(on)}$ Test for FETs

$r_{ds(on)}$ describes the bulk resistance of the channel in series with the drain and source. Channel resistance is particularly important for switching and chopper circuits since it affects the switching speed and determines the output level. Channel resistance is also described as $r_{d(on)}$, $r_{DS}$, and $r_{ds}$, depending upon the data sheet.

Figure 9-30 shows a typical circuit for test of channel resistance in a JFET. Both gates should be tied together, and all terminals should be at the same (zero) d-c voltage. The a-c voltage should be low enough so that there is no pinch-off in the channel. IGFETs can be tested with essentially the same circuit. However, the gate should be biased so that the IGFET is operating in the enhancement mode.

### 9-15.6. Switching Time Tests for FETs

Switching time characteristics are often specified on FET data sheets. These include $t_{d1}$ (delay time), $t_{d2}$ or $t_{ds}$ (storage time), $t_r$ (rise time), $t_f$ (full time). These characteristics must be measured with a pulse source and a multiple-trace oscilloscope, as there are similar characteristics for conventional junction transistors.

Figure 9-31 shows typical circuits for switching time tests of an FET. The procedures are identical to those for conventional junction transistors, described in Section 9-13.

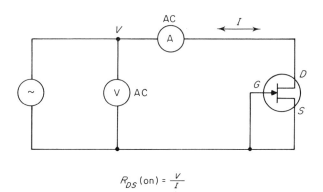

$$R_{DS}(\text{on}) = \frac{V}{I}$$

**Fig. 9-30.** Test circuit for JFET channel resistance.

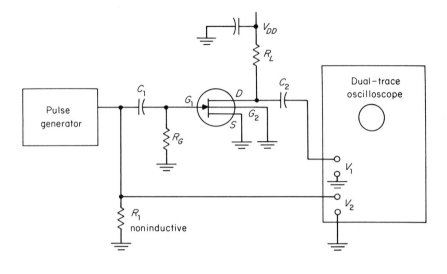

$R_G$ = 1 megohm

$R_L$ = some value that will cause negligible d-c
drop at $I_{DSS}$

$R_1$ = 50 ohms typical

**Fig. 9-31. Typical test circuit for measurement of FET switching characteristics.**

## NOTE

Two other characteristics, $C_{d(\text{sub})}$ and NF, are shown for FETs in Table 9-1. $C_{d(\text{sub})}$ (drain-substrate junction capacitance) is an important characteristic for IGFETs used in switching circuits. This is because $C_{d(\text{sub})}$ appears in parallel with the load in a switching circuit and must be charged and discharged between the two logic levels. NF (noise figure) is usually specified at a particular frequency, since NF will vary with frequency. No special circuit is required for test of either characteristic. However, drain voltage and current, as well as ambient temperature, input resistance, and frequency, should be specified for NF.

## 9-16. Controlled Rectifier Tests

The simplest and most comprehensive test for a controlled rectifier is to operate the unit in a circuit that simulates actual circuit conditions (ac and an appropriate load at the anode, ac or pulse at the gate) and then measure the resulting conduction angle on a dual-trace oscilloscope. The trigger and anode voltages, as well as the load current, can be adjusted to normal (or abnormal) dynamic operating conditions, and the results noted. For example, the trigger voltage can be adjusted so that it is over the supposed minimum and maximum trigger levels. Or the trigger can be removed, and the anode voltage raised to the actual breakover. The conduction angle method should test all important characteristics of a controlled rectifier, except for turn-on, turn-off, and rate-of-rise.

### 9-16.1. Conduction Angle Test

As shown in Fig. 9-32, one trace of a dual-trace oscilloscope displays the anode current, and the other trace displays the trigger voltage. Both traces are voltage calibrated. The anode load current is measured through a 1-ohm noninductive resistor. The voltage developed across this resistor is equal to current (3-V indication equals 3 amperes). The trigger voltage is read out directly on the other oscilloscope trace. Note that a diode has been placed in the trigger circuit to provide a pulsating d-c trigger. This can be removed if desired. Since the trigger is synchronized with anode

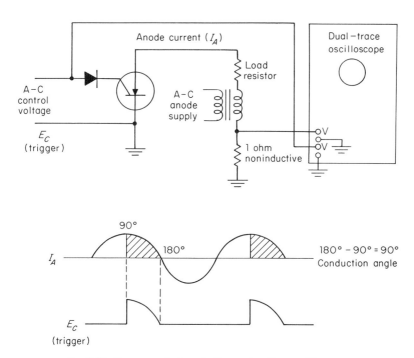

**Fig. 9-32.** Measurement of conduction angle with an oscilloscope.

current, the portion of the trigger cycle in which anode current flows is the conduction angle.

1. Connect the equipment as shown in Fig. 9-32.

2. Place the oscilloscope in operation as described in the instruction manual. Switch on the internal recurrent sweep. Set the sweep selector and sync selector to internal.

3. Apply power to the controlled rectifier. Adjust the trigger voltage, anode voltage, and anode current to the desired levels. Anode voltage can be measured by temporarily moving the oscilloscope probe (normally connected to measure gate voltage) to the anode.

4. Adjust the oscilloscope sweep frequency and sync controls to produce two or three stationary cycles of each wave on the screen.

5. On the basis of one conduction pulse equalling 180°, determine the angle of anode current flow by reference to the trigger voltage trace. For example, in the display of Fig. 9-32, anode current starts to flow at 90° and stops at 180°, giving a conduction angle of 90°.

### NOTE

If the unit under test is a Triac (or similar device), there will be a conduction display on both half cycles.

6. To find the minimum or maximum required *trigger level,* vary the trigger voltage from zero across the level of the trigger voltage when anode conduction starts.

7. To find the *breakdown voltage,* remove the trigger voltage and move the oscilloscope probe to the anode. Increase the anode voltage until conduction starts and note the anode voltage level.

### 9-16.2. Rate-of-rise Tests

When a rapidly rising voltage is applied to the anode of a controlled rectifier, the anode may start to conduct, even though there is no trigger, and the breakdown voltage is not reached. This condition is known as *rate effect, dV/dT effect,* or (sometimes) *dI/dT effect.* The letters $dV/dT$ signify a difference in voltage for a given difference in time. The letters $dI/dT$ signify a difference in current for a given difference in time.

The condition can be tested by a technique known as the *exponential waveform method.* As shown in Fig. 9-33, a large capacitor $C_1$ is charged to the full voltage rating of the controlled rectifier under test. Capacitor $C_1$ is then discharged through a variable time-constant network ($R_2$ and $C_2$).

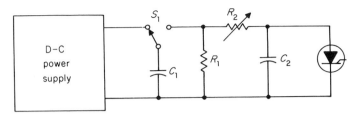

Fig. 9-33. Basic test circuit for rate-of-rise (dV/dT).

This is repeated with smaller time constants (higher $dV/dT$) until the unit under test is turned on by the fast $dV/dT$.

The critical rate, which causes firing, is defined as:

$$dV/dT = \frac{0.632 \times \text{anode voltage}}{R_2 \times C_2}$$

This equation describes the *average slope* of the essentially linear rise portion of the applied voltage (see Fig. 9-34).

In practical test circuits there are two major conditions which determine the value of the circuit components. First, capacitor $C_1$ should be large enough to serve as a constant-voltage source during the discharging of $C_1$ and the charging of $C_2$. Second, capacitor $C_2$ should be much larger than the intrinsic cathode-to-anode capacitance of the unit under test, plus any stray device and device test wiring capacitance. Controlled rectifier junction capacitance is a diminishing function of the applied voltage and has its highest value for zero anode-to-cathode voltage.

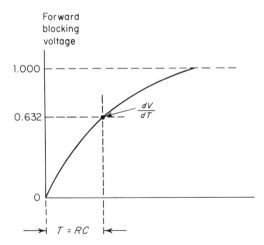

Fig. 9-34. Exponential applied forward voltage and definition of dV/dT.

Typically, for 70-A devices, the junction capacitance has been found to be in the order of 800 pF for zero applied voltage. Generally, 0.5 $\mu$F for $C_1$ and 0.01 $\mu$F for $C_2$ are practical values. Stray inductance and capacitance in the test circuit must be kept to a minimum, especially for measurement of high $dV/dT$ values.

### 9-16.3. Switching Tests for Controlled Rectifiers

Figure 9-35 shows a circuit capable of measuring both turn-on and turn-off (recovery) time of controlled rectifiers. External pulse sources must be provided for the circuit. These pulses are applied to transformers $T_1$ and $T_2$ and serve to turn the unit under test on and off. The pulses can come from any source but should be of the amplitude, duration, and repetition rate that corresponds to the normal operating conditions of the unit under test. When a suitable gate pulse is applied to $T_1$, the unit under test is turned on. Load current can be set by resistor $R_L$. A predetermined time later, the turn-off controlled rectifier is turned on by a pulse, a reverse bias is applied, and the unit under test is turned off.

Any oscilloscope capable of a 10-$\mu$s sweep can be used for viewing both the turn-on and turn-off action. The oscilloscope is connected with the vertical input across the unit under test. Turn-on time is displayed

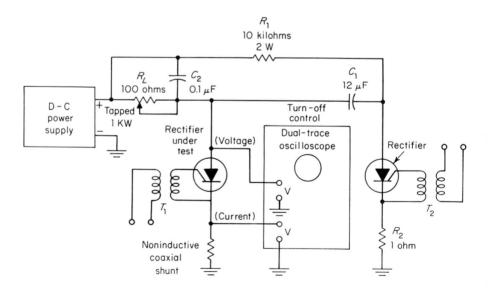

$T_1$ and $T_2$ = secondary windings of pulse transformer

**Fig. 9-35. Test circuit for turn-on and turn-off (recovery) time of controlled rectifiers.**

when the oscilloscope is triggered by the gate pulse applied to the unit under test. Turn-off time is displayed when the oscilloscope is triggered by the gate pulse applied to the turn-off controlled rectifier.

The actual spacing between the turn-on and turn-off pulses is usually not critical. However, a greater spacing will cause increased conduction, heating the junction of the unit under test. Since operation of controlled rectifiers is temperature dependent, the rise in junction temperature must be taken into account for accurate test results. (Both turn-on and turn-off times increase with an increase in junction temperature.)

Figure 9-36 shows turn-on action. Turn-on time is equal to delay time ($t_d$) plus rise time ($t_r$). Following the beginning of the gate pulse, there is a short delay before appreciable load current flows. Delay time is the time from the leading edge of the gate current pulse (beginning of oscilloscope sweep) to the point of 10% load current flow. (Delay time can be decreased by overdriving the gate.)

Rise time ($t_r$) is the time the load current increases from 10% to 90%

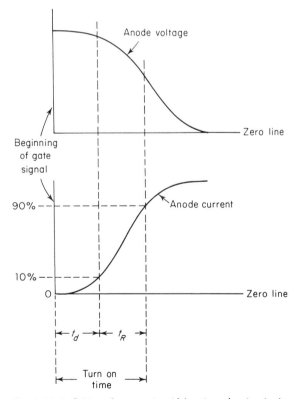

**Fig. 9-36.** Definition of turn-on time (delay time plus rise time).

of its value. Rise time depends upon load inductance, load current amplitude, junction temperature, and to a lesser degree anode voltage. The higher the inductance and load current, the longer the rise time. An increase in anode voltage tends to decrease the rise time. Capacitor $C_2$ (Fig. 9-35) tends to counter the load inductance, thereby lessening the rise time.

By triggering the oscilloscope with the gate pulse applied to the unit under test, the sweep starts at the gate pulse leading edge. Thus, the oscilloscope presentation shows the anode voltage from this point on. In noninductive circuits, when the anode voltage decreases to 90% of initial value, this time is equal to 10% of the load current and is therefore equal to the delay time.

With the oscilloscope set at 10-$\mu$s sweep, each division represents 1 $\mu$s. The delay time is read directly by counting the number of divisions. If the circuit is *noninductive,* the decrease from 90% to 10% of the anode voltage is *approximately* equal to the increase from 10% to 90% of the load current. The time this takes is equal to the rise time. The total time from zero time (start of oscilloscope sweep) to 10% of the anode voltage is the turn-on time. Therefore, turn-on time is determined by counting the number of divisions from the start of the oscilloscope sweep to 90% anode load current.

Figure 9-37 shows the reverse current and reverse recovery (turn-off) action of the unit under test. Turn-off time is the time necessary for the unit under test to turn off *and recover its forward blocking ability.* The reverse recovery time $(t_h)$ is the length of the interval between the time the forward current falls to zero when going reverse and the time it returns to zero from the reverse direction.

In Fig. 9-35, the time available for turn-off action is determined by

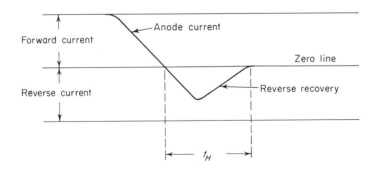

$t_H$ = reverse recovery time

**Fig. 9-37.** Definition of reverse recovery time.

the value of capacitor $C_1$ and resistor $R_2$. Decreasing the value of $C_1$ decreases the time the unit under test is reverse biased. Resistor $R_2$ limits the magnitude of the reverse current. The shapes of the reverse voltage and current pulses are given by a capacitor-resistor discharge. At the end of the reverse pulse, forward voltage is reapplied. Having turned off, the unit under test blocks forward voltage, and no current can flow.

Figure 9-37 shows the reverse current pulse. With the oscilloscope set at 20-$\mu$s sweep, the value of reverse recovery time (turn-off time) can be measured by counting off the divisions from the zero point on the leading edge of the reverse current pulse to the zero point on the trailing edge.

## 9-17. Integrated Circuit Tests

Since an IC is essentially a complete, functioning circuit, it should be so tested. For example, assume that an IC module is used as an amplifier, either differential or operational. Such a module should be submitted to the usual amplifier tests (decibel gain, frequency response, etc.). The procedures for performing such tests on an IC module are the same as for a conventional transistor (or vacuum tube) amplifier.

While the *operating* test procedures for an IC are the same as for conventional circuits, the measurement of static (d-c) voltage applied to the IC is not identical. Most ICs require connection to *both* a *positive and negative* power source. A few ICs can be operated from a single power supply source. Many ICs require equal power supply voltages (such as +9 V and −9 V). However, this is not the case with the example circuit of Fig. 9-38, which requires a +9 V at pin 8 and a −4.2 V at pin 4.

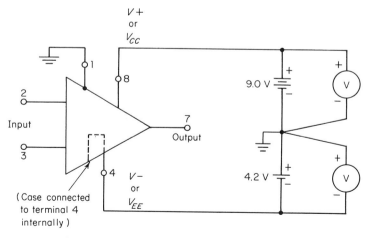

**Fig. 9-38.** Measuring static voltages of integrated circuits.

Unlike most transistor circuits, where it is common to label one power supply lead positive and the other negative without specifying which (if either) is common or ground, it is *necessary that all IC power supply voltages be referenced to a common or ground.*

Manufacturers do not agree on power supply labeling for ICs. For example, the IC of Fig. 9-38 uses $V+$ to indicate the positive voltage and $V-$ to indicate the negative voltage. Another manufacturer might use the symbols $V_{EE}$ and $V_{CC}$ to represent negative and positive respectively. As a result, the IC data sheet should be studied carefully before measuring the power source voltages.

No matter what labeling is used, the IC will usually require two power sources with the positive lead of one and the negative lead of the other tied to ground. Each voltage must be measured separately, as shown in Fig. 9-38.

Note that the IC case (such as a TO-5) of the Fig. 9-38 circuit is connected to pin 4. This is typical for most ICs. Therefore, the case will be below ground (or "hot") by 4.2 V.

CHAPTER **10**

# Microwave Measurements

Because of the high frequencies involved, microwave measurements require specialized test equipment. As stated in the Preface, the instruction manuals supplied with such specialized test equipment generally describe both the operation and application in great detail. Therefore, this chapter is devoted to the basic principles involved with microwave measurements. A thorough study of this chapter will familiarize the reader with microwave measurement procedures in common use. It is assumed that the reader is already familiar with microwave principles, such as the operating theory of waveguides, resonant cavities, klystrons, magnetrons, etc.

## 10-1. Microwave Power Measurement

The current and voltage in a microwave circuit are complex in nature and difficult to evaluate in terms of their ability to do work. On the other hand, power is a real quantity that can be measured and easily related to circuit performance. Also, in a loss-free line, microwave power remains constant with position of measurement (unlike voltage and current levels along the line). Therefore, power is one of the basic measurements made at microwave frequencies.

### 10-1.1. Bolometric Power Measurements

Below 10 mW, power is usually measured with bolometers (temperature-sensitive resistive elements) in conjunction with a balanced bridge. The basic bolometer is a thin wire placed within the RF energy field. Changes in temperature (caused by changes in RF field signal strength) cause a corresponding change in wire resistance.

There are two general types of bolometers: *thermistors* (Fig. 10-1),

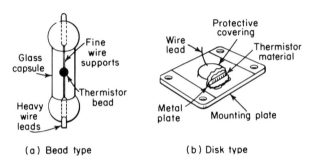

(a) Bead type                    (b) Disk type

**Fig. 10-1.** Typical thermistor.

whose resistance decreases with temperature increases (negative temperature coefficient), and *barretters* (Fig. 10-2), which have a positive temperature coefficient. Thermistors are more commonly used since they are more rugged, both physically and electrically, than the barretters.

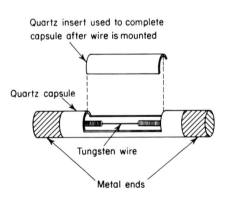

**Fig. 10-2.** Typical barretter.

Bolometer elements are mounted in devices that ideally present a perfect impedance match to microwave transmission lines, either coaxial or waveguide (Fig. 10-3). Such devices, appropriately termed *bolometer mounts,* allow a "bias" connection to the bolometer element, as well as a proper entry point for the RF signal. Waveguide bolometer mounts are the most common for microwave measurements.

In use, the bolometer is connected as one leg of a Wheatstone bridge (or some modification thereof) through the bias connection, and an excita-

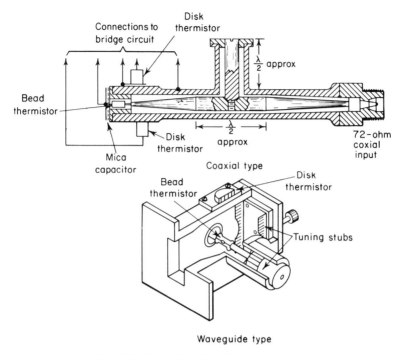

Fig. 10-3. Typical thermistor (bolometer) mounts.

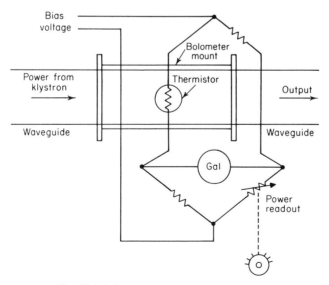

Fig. 10-4. Bolometer power measurement bridge.

tion voltage is applied to the bridge (Fig. 10-4). The d-c or low-frequency a-c bridge excitation serves as the bolometer-element bias power that affects the bolometer's resistance, so that the bridge is essentially balanced. When the unknown RF energy (supplied by a klystron in this case) is applied to the bolometer, the resulting temperature rise causes the element's resistance to change, tending to unbalance the bridge. By withdrawing a like amount of d-c or a-c bias power from the element, the bridge may be returned to balance, and the amount of bias power removed can be converted to a power readout.

### 10-1.2. Calorimetric Power Measurements

When the power to be measured is about approximately 10 mW, bolometer elements are not used directly, since the barretter or thermistor could be damaged. (However, there are thermistors capable of dissipating 50 mW without damage.) Instead, the basic bolometer circuit is used with some form of calorimetric measurement system. The basic difference between the two systems is that in a noncalorimetric meter the temperature sensor is placed directly in the path of the RF power, whereas in a calorimetric meter the power is used to heat a resistance, and the resultant heat is transferred to the temperature sensor.

Calorimetric power meters fall into two classes: dry and fluid. Dry calorimeters depend on a static thermal path between the dissipative load or resistor and the temperature sensor (such as mounting both dissipating resistor and sensor in the same heat sink). This arrangement often requires several minutes for both resistor and sensor to stabilize and is time consuming. Fluid calorimeters such as shown in Fig. 10-5 use a moving stream of oil to transfer heat quickly to the sensing element. An amplifier-feedback arrangement, in conjunction with the series oil-flow system, reduces measurement time to a few seconds for full-scale response.

### 10-1.3. Peak Power Measurements

It is often necessary to measure peak power of pulses used in microwave work. This can be done by using some form of bolometer. However, it is usually more practical to use some form of *video comparator* circuit. In such circuits, a known d-c voltage (across a known impedance) is adjusted to a level equal to the pulse being measured.

In a practical peak power measurement circuit, such as shown in Fig. 10-6, the RF signal is picked up by an RF probe and applied to a diode detector. The demodulated diode output and the output of a d-c reference supply are simultaneously fed to the input of an oscilloscope through a "chopper." The d-c reference voltage is adjusted so that it is exactly equal

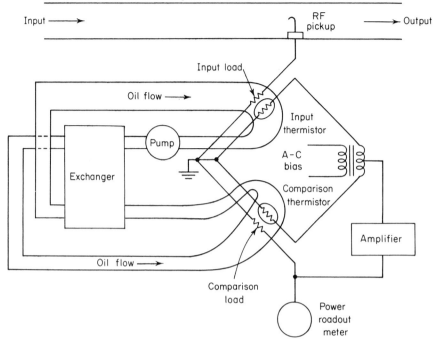

Fig. 10-5. Fluid calorimetric power meter.

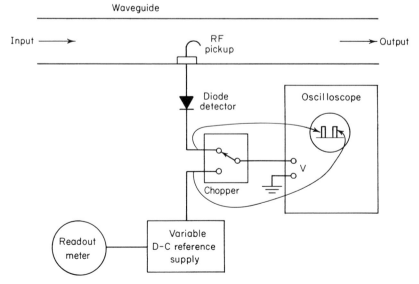

Fig. 10-6. Peak power measurement circuit (video comparator).

225

to the peak value of the demodulated pulse. The level of the required d-c reference voltage is then indicated on the panel meter, calibrated to read RF power.

## 10-2. Microwave Measurement Devices

Although there are many devices used in microwave measurements, those most commonly used are the RF probe [which may include a bolometer (previously discussed) or a diode detector], a slotted line, a directional coupler, and a dummy load. The following paragraphs provide a brief description of each device.

### 10-2.1. RF Probe

Figure 10-7 shows the construction of a typical RF probe, mounted on a slotted line. The probe wire is adjustable for depth of penetration (coupling), so that the amount of RF pickup can be controlled. In practice, the coupling is kept at a minimum to reduce distortion of the fields inside the transmission medium.

The bolometer element shown in Fig. 10-7 may be replaced by a diode detector. When the switch is set to "bolometer," a bias current is applied. The bias current is removed when the switch is set to "crystal." Either way,

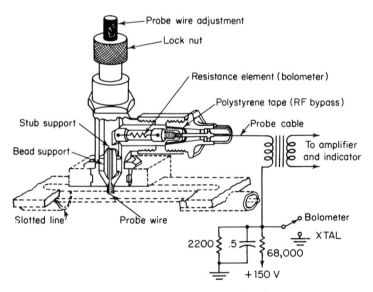

**Fig. 10-7.** Typical RF probe and slotted line.

the RF probe output is coupled to an indicating circuit through a transformer (omitted with some probes). The indicating meter may be a microammeter located directly in the probe cable circuit, or it may be an ammeter preceded by a d-c amplifier to which the probe voltage is fed. Usually, if amplification (greater sensitivity) is desired, the microwave signal source is amplitude modulated with an AF square wave, so that the detector output can in turn be amplified by normal audio circuitry.

### 10-2.2. Slotted Line

As shown in Fig. 10-7, the slotted line is a coaxial or waveguide section of transmission line, with a longitudinal slot cut into its outer shell to permit the insertion of a probe. The slot is generally at least one wavelength long and is narrow enough to cause very little loss due to radiation leakage. Through this slot a probe can be placed in the electromagnetic field inside the section and moved up and down to explore the voltage field. The most common use for a slotted line is to measure standing wave ratios (SWR) in microwave equipment.

### 10-2.3. Diode Detector

The diode detectors used in microwave RF probes are essentially the same as those used at lower frequencies. Their function is to convert RF power into d-c voltage. This voltage can then be monitored with a d-c voltmeter and calibrated in terms of power or whatever other value is desired. Diode probes are often used with VTVMs or electronic voltmeters where there is some amplification involved. When used for microwave work, diode detectors must be designed in conjunction with the RF probe to provide minimum capacitance and maximum input resistance. A high capacitance could produce considerable capacitive reactance in a microwave circuit. On the other hand, maximum input resistance keeps the current at an absolute minimum.

### 10-2.4. Directional Coupler

A directional coupler functions to pass RF signals in one direction only and to sample a portion of the RF signals (Fig. 10-8). The detector in the directional coupler produces a d-c voltage in proportion to the RF signal passing through the coupler.

Directional couplers serve as stable, accurate, and relatively broad-band coupling devices that can be inserted into a transmission line so as to measure either incident or reflected power (depending upon the direction of the coupling).

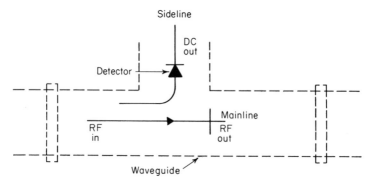

**Fig. 10-8.** Directional coupler used in microwave waveguides.

A measure of the coupling of a directional coupler is a ratio of the power into the main-line terminal to the power out of the side-line terminal. Coupling in decibels = 10 log (power in/power out).

Insertion losses of a directional coupler (or any similar device) are a measure of the power lost because of the insertion of the device into a microwave transmission system. Although this could be measured in terms of absolute power units, it is universally expressed in terms of power-ratio units (decibels, etc.).

For attenuation (insertion-loss) measurements in microwave equipment, the decibel is defined as

$$1 \text{ dB} = 10 \log \frac{P_1}{P_2}$$

where $P_1$ is the power absorbed at the load without the device in the line and $P_2$ is the power absorbed with the device in the line.

### 10-2.5. Dummy Load

A dummy load is used to prevent the radiation of power that could cause interference. Dummy loads are required for service and test procedures on microwave equipment, as they are for lower-frequency units. However, a microwave dummy load must be of special design so as not to create standing waves. A conventional wire-wound resistor, for example, will create a high inductive reactance at microwave frequencies. This reactance will cause a severe mismatch between the waveguide and the dummy load, resulting in standing waves, power loss, etc. If such a dummy load were connected directly to a microwave transmitter output, it might even affect the transmitter tuning. At the normal operating frequency, a microwave antenna is almost pure resistive. The dummy load that replaces such an antenna must also present almost pure resistance with an absolute minimum of inductive reactance.

Many times *sliding or variable* dummy loads are used when testing microwave components in the laboratory. This permits the dummy load to be matched to the exact impedance of the waveguide. *Sliding shorts* are also used to terminate some waveguide test setups. A nonadjustable short can produce standing waves at microwave frequencies.

## 10-3. Fixed-frequency Versus
## Swept-frequency Measurements

Both fixed-frequency and swept-frequency techniques are used for the measurement of power, attenuation, frequency, impedance, etc., in microwave equipment.

*Fixed-frequency* techniques offer the highest precision attainable for individual measurements, because the small inherent mismatch errors that must be tolerated on a broad frequency-sweep basis may be individually tuned out. Consequently, fixed-frequency techniques are widely used in "standard" measurements and in applications in which the system under test is operating either at a single frequency or within a very narrow band.

*Swept-frequency* techniques are used to obtain measurements quickly and easily over a range of frequencies. Important parameters such as SWR, directivity, attenuation, noise figure, etc., can be determined on a swept-frequency basis, and the user can quickly determine whether there is a narrow-band condition, such as a resonance, in the device being tested.

## 10-4. Microwave Impedance Measurements

The most important consideration in microwave impedance measurements is in *impedance matching* a load to its source. This is usually more important than actual impedance value. If the load and source are mismatched, part of the power will be reflected back along the transmission line toward the source. This reflection not only prevents maximum power transfer but also can be responsible for erroneous measurements and can easily cause circuit damage in high-power applications. The power reflected from the load interferes with the incident (forward) power, causing standing waves of voltage and current along the line. The ratio of standing-wave maximums to minimums is directly related to the impedance mismatch of the load. The standing-wave ratio therefore provides a valuable means of determining impedance and mismatch. The subject of standing-wave ratios is discussed further in Chapter 11.

### 10-4.1. Fixed-frequency Impedance Measurement

Standing-wave ratio can be measured directly at fixed frequencies using a slotted line. In use, the slotted line is placed immediately ahead of the load, as shown in Fig. 10-9, and the source is adjusted for some fixed

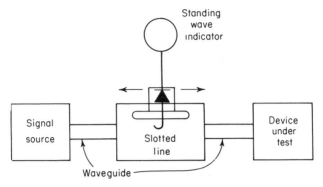

**Fig. 10-9.** Basic slotted-line test connections.

amplitude-modulation frequency (usually 1 kHz) at the desired micro-wave frequency. The slotted-line probe is loosely coupled to the RF field in the line, thus sensing relative amplitudes in the standing-wave pattern as the probe is moved along the line. The ratio of maximum to minimum voltage is then read directly on the standing-wave indicator.

For example, assume that the forward signal was 10 V and that the reflected signal was 2 V. This would produce standing waves with 12-V maximums (where the forward and reflected voltages were in-phase and adding) and 8-V minimums (where the forward and reflected voltages were out-of-phase and subtracting). This would mean a voltage standing-wave ratio (VSWR) of $12/8 = 1.5$.

Because the probe must not be allowed to extract any appreciable power from the slotted line, high sensitivity and low noise are required in the detector and indicator. To this end, the indicator is sharply tuned to the modulation frequency of the source, thereby reducing noise and allow-ing the use of a high-gain audio amplifier and voltmeter circuit.

Other considerations relative to accurate slotted-line measurements in-clude elimination of harmonics from the source prior to entering the slotted line, low-frequency modulation in the source, and a low residual standing-wave ratio in the slotted line itself.

### 10-4.2. Swept-frequency Impedance Measurement

Figure 10-10 shows the basic circuit for swept-frequency impedance measurement of microwave equipment using the *reflection* system. The basic idea of this method is to hold the forward voltage constant and measure *return loss* of a given load rather than the direct ratio of forward

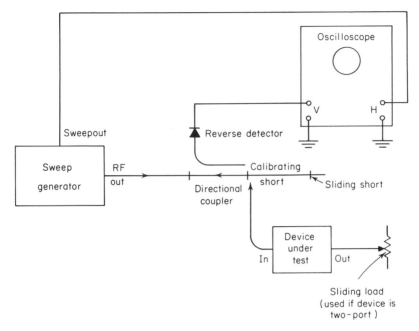

**Fig. 10-10.** Microwave sweep-frequency impedance measurement.

and reflected signal voltages. This technique eliminates the need for a slotted line and amplitude modulation of the generator signal.

In operation, a calibrating short is placed on the waveguide output to set up a 100% reflection to the reverse-coupler detector as the generator sweeps the band. Since this reflection is 100%, it is equal to the forward signal. The output from the reverse detector is applied to the oscilloscope and is adjusted to some convenient reference point representing the forward signal level. When the reference is established, the short is replaced by the unknown load. The amplitude decrease on the oscilloscope indicates the *load return loss*.

For example, assume that the forward voltage is 10 V and the reflected voltage is 2 V (as in the case of the fixed-frequency measurement, Section 10-4.1). With the calibrating short in place, the forward (and reflected) signal would be 10 V. The oscilloscope is then adjusted to some convenient level, such as ten vertical scale divisions. After the short is replaced by the unknown load, the oscilloscope indication would drop to two divisions (indicating a 2-V reflected signal). The difference between the forward and reflected voltages can then be converted to VSWR, if desired.

The term *reflection coefficient* is often used in microwave work instead of VSWR.

$$\text{Reflection coefficient} = \frac{\text{reflected voltage}}{\text{forward voltage}}$$

For example, using 10 V forward and 2 V reflected, the reflection coefficient is $\frac{2}{10} = 0.2$.

Reflection coefficient can be converted into VSWR by dividing (1 + reflection coefficient) by (1 − reflection coefficient). For example, using the 0.2 reflection coefficient, the SWR is

$$\frac{1 + 0.2}{1 - 0.2} = \frac{1.2}{0.8} = 1.5 \text{ VSWR}$$

VSWR can be converted into the reflection coefficient by dividing (VSWR − 1) by (VSWR + 1). For example, using the 1.5 VSWR, the reflection coefficient is

$$\frac{1.5 - 1}{1.5 + 1} = \frac{0.5}{2.5} = 0.2 \text{ reflection coefficient}$$

## 10-5. Microwave Attenuation Measurements

It is possible to measure the attenuation of microwave components at fixed frequencies. However, the swept-frequency techniques are becoming more popular for reasons discussed in Section 10-3. In addition, it is possible to use an *X-Y* recorder instead of an oscilloscope with swept-frequency measurements. An *X-Y* recorder requires a sweep voltage or trigger voltage (supplied by the sweep generator) and provides a permanent record on chart paper instead of an oscilloscope display. The *X-Y* recorder/sweep generator combination can also be used for other microwave measurements, such as SWR, impedance, power, frequency, etc., if desired.

The basic setup for making swept-frequency attenuator measurements is shown in Fig. 10-11. A portion of the sweep generator output signal is removed by the directional coupler and fed back to the power output level-control circuits of the sweep generator. This closed loop provides a means of maintaining the sweep generator output constant. Any variation in sweep generator output amplitude will cause an error. The second directional coupler is arranged so that the reflected channel of the impedance test setup (Fig. 10-10) now becomes the transmission channel.

The system is first calibrated by placing a length of loss-free waveguide between the source and the *X-Y* recorder detector. A reading is obtained on the *X-Y* recorder with the section of loss-free waveguide in the circuit. The *X-Y* recorder circuits are adjusted to some specific refer-

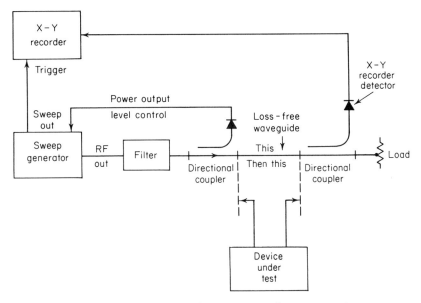

**Fig. 10-11.** Microwave sweep-frequency attenuation measurement.

ence level representing zero attenuation. The loss-free waveguide is then removed, and the device to be tested is connected in its place. The resulting signal will then be detected, and the value will be displayed on the *X-Y* recorder. Usually, the *X-Y* recorder indications are related to voltage. Therefore, the ratio between the two voltages can be converted into a loss figure expressed in decibels, if desired. Refer to Chapter 7 for information on decibel measurements.

## 10-6. Microwave Frequency Measurements

There are four devices used in microwave frequency measurements: the Lecher line, the LC meter, the cavity meter, and some form of heterodyne or "beat frequency" meter.

The *Lecher line* is used primarily in *experimental* microwave work to measure wavelength. A typical Lecher line is simply a two-wire line with a sliding shorting bar (or meter with sliding contacts) across the two lines (see Fig. 10-12). In Fig. 10-12a, the current-indicating device (probably a thermocouple ammeter) is moved along the line to measure points of maximum and minimum current. The distance between successive maximums or between successive minimums is equal to a half wavelength. (This can be converted into frequency as discussed in Chapter 6.) In Fig. 10-12b, the shorting bar is moved until the line becomes resonant at the

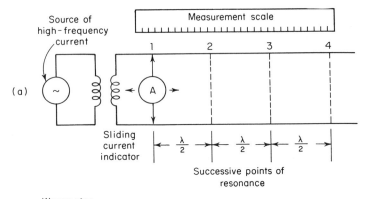

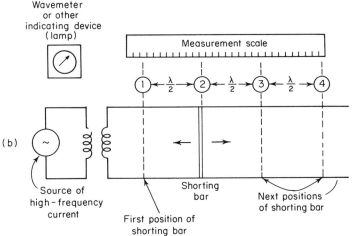

**Fig. 10-12.** Lecher lines for microwave frequency measurements.

generator frequency, as indicated by the wavemeter. The actual wavelength can be measured and the frequency calculated from this measurement. Lecher lines are used at frequencies up to about 3 GHz.

The *LC meter* is simply a resonant circuit with variable inductance or capacitance. LC meters are usually not accurate above about 1.5 GHz and present several problems in their use. For example, the LC meter must be very lightly coupled to the circuit under test. Otherwise, inaccuracies will occur, and the meter indicating device could burn out. For these reasons, the LC meter has been replaced by the cavity wavemeter.

The *cavity wavemeter* is typical of passive microwave frequency measurement devices (see Fig. 10-13). These wavemeters are two-port devices that absorb part of the input power in a tunable resonant cavity. When the cavity is tuned to resonate at the input frequency, a dip occurs in the output power level that is measured by a meter or oscilloscope. Frequency is then read directly on a calibrated dial driven by the cavity tuning mechanism. The accuracy of cavity wavemeters depends upon the cavity Q, tuning dial calibration, backlash, and the effects of temperature and humidity variations. Present-day cavity wavemeters are available with accuracies of a few parts in $10^4$, at frequencies up to 40 GHz.

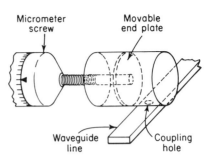

Wavemeters are often used in laboratories to check signal generator accuracy. The wavemeters are calibrated against some form of heterodyne meter at routine intervals. The wavemeters are then used to check signal generator frequency just prior

**Fig. 10-13.** Typical cavity wavemeter for microwave frequency measurement.

to use or are incorporated into the test circuit to provide continuous monitoring of the signal generator frequency.

The *heterodyne frequency* meters used for microwave work are similar to those of lower frequency (refer to Chapter 6). In basic operation, a signal of known frequency is mixed with a signal of unknown frequency. The unknown signal frequency is adjusted until there is an indication that it is at the same frequency as the known signal (or multiple thereof).

In microwave work, the heterodyne frequency measurement system most often involves a sweep generator, transfer oscillator, and electronic counter. Such systems provide frequency measurements up to about 40 GHz, with accuracies up to a few parts in $10^8$.

Figure 10-14 shows a typical heterodyne measurement setup used to calibrate cavity wavemeters. In operation, the sweep generator is set for a leveled, rapidly sweeping output across a small segment of the cavity wavemeter's band, with the main power being fed through the wavemeter to a diode detector. The detected wavemeter dip is fed to the vertical input of the oscilloscope. A portion of the sweep generator output is applied to the input of the transfer oscillator (also known as a frequency converter). The transfer oscillator is essentially a crystal-controlled oscillator of great accuracy and a mixer circuit. (Usually, some form of crystal mixer is used.) Harmonics of the transfer oscillator mix with the sweeping input signal, producing heterodyne signals (often called "birdies" or "pips") at intervals equal to the transfer oscillator fundamental frequency. The actual transfer oscillator frequency is read out by the electronic counter.

For example, if the transfer oscillator frequency is 200 MHz and the

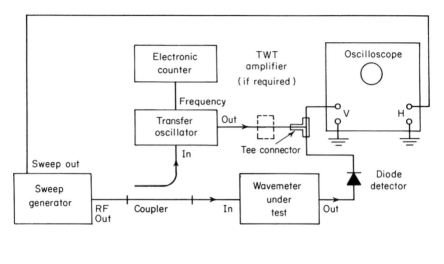

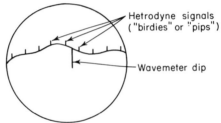

**Fig. 10-14.** Microwave sweep-frequency measurement of wavemeters.

input frequency is sweeping from 10 to 20 GHz (or any other band), "birdies" will appear every 200 MHz throughout the swept band as shown in Fig. 10-14. With such a display on the oscilloscope, the wavemeter is adjusted for coincidence of its dip and that of the reference "birdies," noting the accuracy of the cavity wavemeter dial calibration at each point.

As shown in Fig. 10-14, the output of a typical transfer oscillator can be used directly up to about 18 or 20 GHz. At higher frequencies, it is usually necessary to use some form of amplification (such as a traveling wave tube) for the transfer oscillator output.

# CHAPTER 11

# Antenna and Transmission
# Line Measurements

## 11-1. Antenna Length and Resonance
## Measurements

Most antennas are cut to a length related to the wavelength of the signals being transmitted or received. Generally, antennas are cut to one-half wavelength (or one-quarter wavelength) of the center operating frequency.

The electrical length of an antenna is always greater than the physical length, owing to capacitance and end effects. Therefore, two sets of calculations are required: one for electrical length and one for physical length.

The calculations for antenna length and resonant frequency are shown in Fig. 11-1.

### 11-1.1. Practical Resonance Measurements
### for Antennas

With a short antenna it is possible to measure the exact physical length and find the electrical length (and hence the resonant frequency) using the equations of Fig. 11-1. Obviously this is not practical for long antennas. Also, the exact resonant frequency (electrical length) may still be in doubt for short antennas due to the uncertain $K$ factor of Fig. 11-1. Therefore,

237

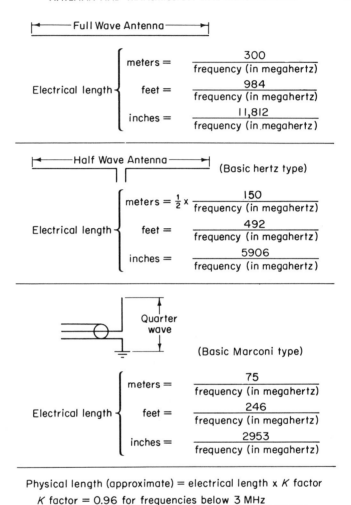

**Fig. 11-1.** Calculations for antenna length.

for practical purposes the electrical length and resonant frequency of an antenna should be determined electrically.

There are three practical methods for determining antenna resonant frequency: dip adapter circuit, antenna ammeter, and wavemeter.

The *dip adapter* (described in previous chapters) can be used to measure resonant frequency of both grounded and ungrounded antennas. The basic technique is to couple the adapter to the antenna as if the antenna were a resonant circuit, tune for a dip, and read the resonant frequency. However, there are certain precautions to be observed.

The measurement can be made as a conventional resonant circuit provided the antenna is accessible allowing the dip adapter to be coupled directly to the antenna elements. If the antenna is a simple grounded element (no matching problems between antenna and transmission line), the adapter can be coupled to the transmission line. However, if the antenna is fed by a coaxial line or any system in which the line is matched to the antenna, the adapter coil must be coupled directly to the antenna elements. No matter how carefully the antenna and lead-in are matched, there will be some mismatch, at least over a range of frequencies. This means that there will be two reactances (or impedances) that will interact to produce extraneous resonances. If resonance is measured under such conditions, a dip will be found at the correct antenna frequency (plus harmonics) and another dip (plus harmonics) at the extraneous frequency. It can be very confusing to tell them apart. Also, antenna resonance measurements should be made with the antenna in the actual operating position. The nearness of directive or reflective elements, as well as the height, will affect antenna characteristics and possibly change resonant frequency.

The dip adapter procedure for grounded antennas is as follows.

1. Couple the dip adapter to the antenna tuner if the antenna is used for the transmission line and the feed line is tuned.

2. If the antenna is untuned or it is not practical to couple to the antenna tuner, disconnect the antenna lead-in and couple the lead-in to the adapter through a pick-up coil as shown in Fig. 11-2.

3. Set the generator to its lowest frequency. Adjust the signal generator output for a convenient reading on the adapter meter.

4. Slowly increase the generator frequency, observing the meter for a dip indication. Tune for the bottom of the dip.

5. Note the frequency at which the first (lowest frequency) dip occurs. This should be the primary resonant frequency of the antenna. As the signal generator frequency is increased, additional dips should be noted. These are harmonics and should be multiples of the primary resonant frequency. Check two or three of these frequencies to be sure that they *are* harmonics. Then go back to the lowest frequency dip to ensure that the lowest frequency is the primary resonant frequency.

The dip adapter procedure for ungrounded antennas is as follows.

1. Disconnect the antenna lead-in or feed line from the antenna.

2. If the antenna is center fed, short across the feed point with a piece of wire.

3. Couple the dip adapter coil directly to the antenna. Either capaci-

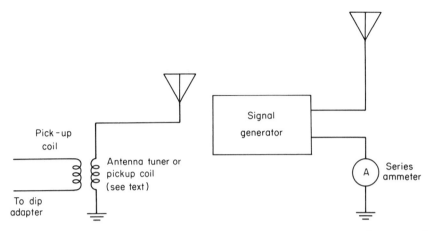

**Fig. 11-2.** Measuring resonant frequency of simple grounded antenna with dip adapter circuit.

**Fig. 11-3.** Measuring resonant frequency of antenna with series ammeter.

tive or inductive coupling can be used. Usually, the best results will be obtained using inductive coupling at a maximum current (low-impedance) point. For example, the maximum current point occurs at the center of a half-wave antenna. If capacitive coupling is used, make the measurement at a minimum current (high-impedance) point.

4. Starting at the lowest signal generator frequency and working upward, tune the signal generator for a dip on the meter as described for grounded antennas. The lowest frequency dip is the primary resonant frequency.

A *series ammeter* can also be used to find the resonant frequency of an antenna. The basic circuit is shown in Fig. 11-3, and the procedure is quite simple. The signal generator is tuned for a maximum reading on the ammeter, indicating a maximum transfer of energy from the generator into the antenna (as a result of both being at the same frequency). The antenna frequency is then read from the generator dial. The series ammeter method has the advantage of measuring the combined resonant frequency of both the antenna and transmission line. This is most practical, since in normal use the antenna will be operated with the transmission line.

A version of the series ammeter method is often used in transmitters as an indicator for antenna tuning. Most transmitters are crystal controlled and operated at a specific frequency with the *antenna tuned to that frequency*. The electrical length (and consequently the resonant frequency) of the antenna is varied by a reactance in series with the lead-in, as shown in Fig. 11-4. The reactance can be a variable capacitor or variable inductance. With such an arrangement, the transmitter is tuned to its operat-

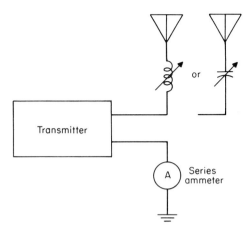

**Fig. 11-4.** Tuning an antenna to resonant frequency of associated transmitter using series ammeter method.

ing frequency, and then the antenna is tuned to that frequency as indicated by a maximum reading on the series ammeter.

The series ammeter method has certain drawbacks, one being the operating frequency limit of the series ammeter. Another is the fact that the series ammeter consumes some power in its operation. However, the series ammeter has an advantage in that true antenna power can be calculated (as discussed in Section 11-2.1).

A *wavemeter* can be used to find the resonant frequency of an antenna. The basic wavemeter circuit is shown in Fig. 11-5 and is essentially a tuned resonant circuit, detector, and indicator. A commercial wavemeter has a precision-calibrated tuning dial so that exact frequency can be measured and an amplifier circuit so that weak signals can be measured. When used to measure antenna resonant frequency, the wavemeter is tuned to the approximate resonant frequency of the antenna, and then the signal generator is tuned for a maximum reading on the wavemeter. In this case, the wavemeter resonant circuit is broadly tuned. When used to tune an antenna, the wavemeter is tuned to the transmitter operating frequency, and then the antenna is tuned for a maximum reading on the wavemeter. Not all wavemeters are provided with precision tuning. Such wavemeters serve only as a maximum (or peak) readout device.

## 11-2. Antenna Impedance and Radiated Power

The impedance of an antenna is not constant along the entire length of the antenna. In a typical half-wave antenna as shown in Fig. 11-6, the impedance is minimum at the center and maximum at the ends. In theory,

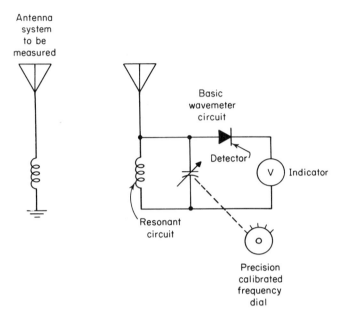

**Fig. 11-5.** Basic wavemeter circuit.

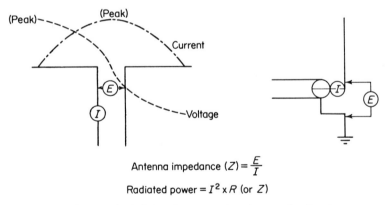

Antenna impedance $(Z) = \dfrac{E}{I}$

Radiated power $= I^2 \times R$ (or $Z$)

**Fig. 11-6.** Theoretical calculations for antenna impedance and radiated power.

the impedance is zero at the center. Since the antenna is fed at some point away from the exact center, there is some impedance for any antenna. A typical half-wave antenna used in commercial transmitters has an impedance of 50 or 72 ohms, while a half-wave TV antenna has 300-ohms impedance.

Antenna impedance is determined using the basic Ohm's law equation $Z = E/I$, with voltage and current being measured at the *antenna feed point*. However, such measurements are not usually made in practical applications.

*Radiation resistance* is a more meaningful term. When the d-c resistance of the antenna is disregarded (antenna d-c resistance is rarely more than an ohm, except in low-frequency long-wire antennas), the antenna impedance can be considered as the radiation resistance. Radiated power can then be determined using the basic Ohm's law equation $P = I^2R$.

### 11-2.1. Practical Impedance and Radiated Power Measurements for Antennas

On those antennas designed to be used with coaxial or twin-lead transmission lines, the antenna and transmission line impedance must be matched. In this case, the *impedance match* between transmission line and antenna is of greater importance than actual impedance value (both antenna and transmission line must be 50 ohms, 72 ohms, 300 ohms, or some other value). The condition of match (or mismatch) between antenna and transmission line can best be measured by the *standing-wave ratio*, or SWR, as is discussed in Section 11-5 (and in Chapter 10).

On those antennas where the lead-in and antenna are considered as one piece with no match or mismatch problems, it is necessary to measure the actual antenna impedance (or radiation resistance) in order to calculate radiated power.

The following procedure can be used to find the impedance and radiated power of any antenna system. However, it should be noted that the impedance and power obtained are for the *complete antenna system* (antenna and transmission line) as seen from the measurement end.

1. Connect the equipment as shown in Fig. 11-7.
2. Set switch $S_1$ to position 1. Adjust the signal generator to the center frequency at which the antenna will be used (or for any desired operating frequency to which the antenna can be tuned). If the antenna is to be used with a transmitter, the transmitter may be substituted for the signal generator.
3. Tune the antenna to the operating frequency by adjusting $L_1$ for a maximum indication on the ammeter. Record the indicated current.
4. If a precision wavemeter is available, verify that the signal generator (or transmitter) and antenna are tuned to the correct frequency.
5. Set switch $S_1$ to position 2. Adjust capacitor $C_1$ for a maximum indication on the ammeter. If the wavemeter is available, verify that the operating frequency has not been changed.

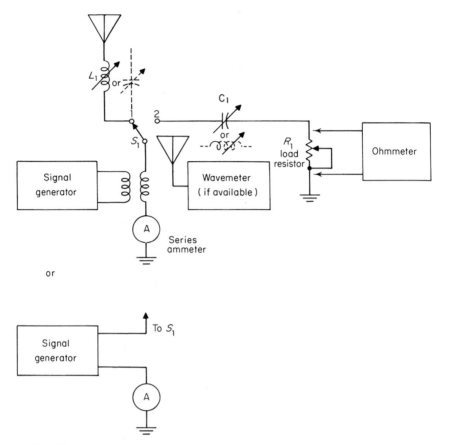

**Fig. 11-7.** Measuring antenna impedance and radiated power using resistance substitution.

## NOTE

If a capacitor has been used to tune the antenna instead of $L_1$, then a precision inductor must be used with resistor $R_1$.

6. Adjust load resistor $R_1$ until the indicated current is the same as the antenna current recorded in Step 3.

7. Remove power from the circuit. Measure the d-c resistance of $R_1$ with an ohmmeter. This resistance is equal to the antenna system impedance (or radiation resistance).

8. Calculate the actual power delivered to the antenna (or radiated power) using

$$\text{radiated power} = I^2 \times R(\text{or } Z)$$

where   $I$ is indicated current in amperes and
        $R$ is radiation resistance (or antenna impedance)

An *alternate method* must be used when the operating frequency is beyond the range of the available ammeter, when no ammeter is present, or when the ammeter will present an excessive load. A precision 1-ohm, noninductive resistor and voltmeter can be used in place of the ammeter, as shown in Fig. 11-8. With a 1-ohm resistor, the indicated voltage will be equal to the current passing through the resistor (and antenna system). Except for the connections, the procedure is identical to the one that requires an ammeter.

## 11-3. Antenna Gain

Antenna gain is a term usually applied to receiving antennas. Such gain is measured by comparing the voltage produced at the terminals of the antenna with that of a thin-wire dipole of the same size, operating at the same frequency, and in the same location. Antenna gain is normally expressed in decibels, since it is essentially a ratio.

Antenna gain is often shown in a gain curve or gain chart. Usually the $0 =$ dB line or reference indicates the gain of the thin-wire dipole to which the antenna is being compared.

Typical gain curves and calculations for antenna gain are shown in Fig. 11-9.

There are several practical methods for measuring antenna gain, as

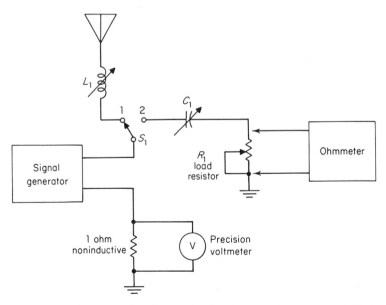

**Fig. 11-8.** Measuring antenna impedance and radiated power using resistance substitution (voltmeter method).

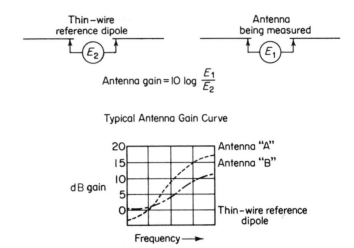

Fig. 11-9. Calculations for antenna voltage gain and typical gain curves.

shown in Fig. 11-10. The simplest is to measure the antenna voltage with a voltmeter and RF probe (Fig. 11-10a). This is usually satisfactory in laboratory work in which a transmitted signal can be directed toward the antenna.

In cases in which the transmitted signal is very weak or it is desired to measure antenna gain at some specific frequency, it is necessary to use a tuning circuit, detector, and amplifier, as shown in Fig. 11-10b. Some commercial wavemeters incorporate such circuits.

Receivers can be used to measure antenna gain. However, the automatic gain control (AGC) or automatic volume control (AVC) circuits of the receiver must be disabled. Also, the receiver must be provided with an output meter that reads in volts or decibels.

The important consideration in making antenna gain measurements is that both the thin-wire dipole and antenna must be tested under identical *conditions* (same frequency, physical location, and test instruments). In some cases, the change of a few inches in physical location will completely change the antenna's gain pattern.

## 11-4. Transmission Line Measurements

Transmission lines are devices used to transfer energy from one unit to another. The most common type of transmission lines found in practical applications are those used to transfer signals to and from antennas. There are two basic types: parallel-wire and coaxial or concentric. In

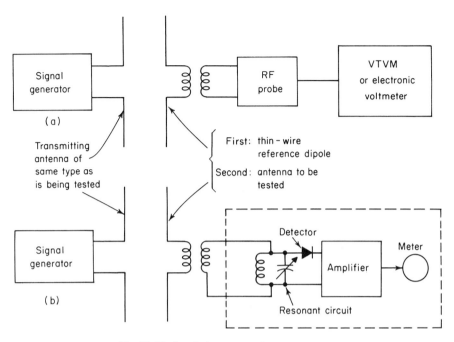

**Fig. 11-10.** Practical antenna gain measurements.

either case, there is capacitance between the lines, as well as inductance set up around the lines. Because of this capacitance and inductance (and resistance of the wire), a transmission line has a characteristic or *surge impedance.* Such impedance is dependent upon the dimensions of the wire or conductor, spacing between the conductors, and dielectric constant of the material between the conductors. Calculations for transmission line impedance are shown in Fig. 11-11.

Any type of transmission line will have some attenuation of the signals passing through it. The attenuation is inversely proportional to the impedance and directly proportional to the direct current resistance. Calculations for transmission line attenuation are also shown in Fig. 11-11.

In some applications it is convenient to obtain the radio-frequency resistance of transmission lines. Such resistance is dependent upon frequency because of the "skin effect" whereby radio-frequency signals tend to travel only on the outside of the conductor.

Calculations for transmission line radio-frequency resistance are also shown in Fig. 11-11.

As in the case of antennas, transmission lines can be cut to wavelengths of signals passing through them. The same basic equations (Section 11-1 and Fig. 11-1) can be used to determine the electrical wavelength. The

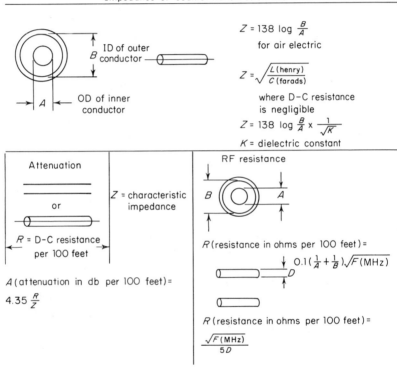

Impedance of parallel-wire transmission line

$Z = 276 \log \frac{2B}{A}$ for air dielectric or

$Z = \sqrt{\dfrac{L\,(\text{henry})}{C\,(\text{farads})}}$

where D-C resistance is negligible

Impedance of coaxial transmission line

$Z = 138 \log \dfrac{B}{A}$ for air electric

$Z = \sqrt{\dfrac{L\,(\text{henry})}{C\,(\text{farads})}}$

where D-C resistance is negligible

$Z = 138 \log \dfrac{B}{A} \times \dfrac{1}{\sqrt{K}}$

$K$ = dielectric constant

Attenuation

$Z$ = characteristic impedance

or

$R$ = D-C resistance per 100 feet

$A$ (attenuation in db per 100 feet) =

$4.35 \dfrac{R}{Z}$

RF resistance

$R$ (resistance in ohms per 100 feet) =

$0.1 \left( \dfrac{1}{A} + \dfrac{1}{B} \right) \sqrt{F\,(\text{MHz})}$

$R$ (resistance in ohms per 100 feet) =

$\dfrac{\sqrt{F\,(\text{MHz})}}{5D}$

**Fig. 11-11.** Calculations for transmission line impedance, attenuation, and RF resistance.

physical length of the transmission line will be shorter, however, since radio waves travel faster in free space than along a conductor. The ratio of actual velocity along a transmission line versus velocity of the same signals in free space is known as the *velocity factor*. This factor is always less than 1 and usually varies between about 0.6 and 0.97 for typical transmission lines.

To determine the physical length of a transmission line for a given frequency, first determine the electrical wavelength for that frequency, then multiply the result by the velocity factor for that type of transmission line.

Table 11-1 lists the velocity factors for a number of typical transmission lines.

**TABLE 11-1**

Velocity Factors of Transmission Lines

| Type of Line | Velocity Factor |
|---|---|
| Twisted-pair line (rubber dielectric) | 0.6 |
| Two-wire line (plastic dielectric | 0.75 |
| Coaxial line (solid plastic dielectric) | 0.67 |
| Coaxial line (air dielectric) | 0.85 |
| Parallel tubing (air dielectric) | 0.95 |
| Two-wire open line (air dielectric) | 0.97 |

(Free-space propagation velocity is $300 \times 10^6$ m/s, or $984 \times 10^6$ feet/s.)

### 11-4.1. Practical Transmission Line Measurements

In practical applications, it may become necessary to find transmission line impedance and length. In the case of coaxial lines, the impedances are listed in manufacturer's data (and in various tables) for coaxial types. The impedance should be the same for all coaxial cables of the same type without regard to the manufacturer. However, the impedance data may not be readily available. Also, the transmission line impedance may change due to deterioration.

In any event, it is a simple matter to measure transmission line impedance. The sweep frequency method (Chapter 6) can be used to provide impedance measurement over a wide range of frequencies. If sweep equipment is not available, the dip adapter method can be used as described in the following paragraphs.

In addition to measuring transmission line impedance, it is often convenient to measure transmission line length by electrical means, as for instance when a line is installed at some inaccessible location or when a large length of transmission line is coiled. A dip adapter can be used to measure the physical distance from the accessible end to the opposite end. The inaccessible end can be open or shorted.

### 11-4.2. Measuring Transmission Line Impedance with a Dip Adapter

1. Select a variable resistance whose range covers the supposed impedance of the line. The resistance must be *noninductive* (such as a carbon or composition potentiometer).

2. Connect the equipment as shown in Fig. 11-12, but do not connect the variable resistance at this time.

3. Set the signal generator to its lowest frequency. Adjust the signal generator output for a convenient reading of the adapter meter.

4. Slowly increase the signal generator frequency, observing the meter for a dip indication. Tune for the bottom of a dip.

## NOTE

This is the resonant frequency of the transmission line. Additional harmonic dip indications should be found at multiples of this frequency. The lowest or primary resonant frequency is the point at which the transmission line is a *quarter wavelength*.

5. With the signal generator tuned to the bottom of a dip at the primary resonant frequency, connect the variable resistance to the open end of the line.

6. Adjust the variable resistance to the supposed impedance of the line. This should *remove the dip.*

7. If the dip is still present, continue adjustment of the resistance near the supposed impedance until the dip is completely eliminated. Check the settings by removing the resistance. The dip should reappear.

8. Once it has been determined that the line has been terminated in its characteristic impedance, measure the resistance of the potentiometer with an ohmmeter. This resistance is equal to the transmission line impedance.

### 11-4.3. Measuring Transmission Line Length with a Dip Adapter

1. Couple the dip adapter to the accessible end of the line as shown in Fig. 11-12. Do not connect a terminating resistor.

2. Set the signal generator to its lowest frequency. Adjust the signal generator output for a convenient reading of the adapter meter.

3. Slowly increase the generator frequency, observing the meter for a

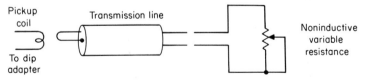

**Fig. 11-12.** Measuring transmission line characteristics with dip adapter circuit.

dip indication. Tune for the bottom of a dip. Continue to increase generator frequency, watching for additional dips.

4. Loosen the coupling until the dips become small and then record
the frequency readings of any two adjacent dips.

5. Using the two dip-indication frequency readings, calculate the line
length using the following equation

$$D = \frac{984 \times K}{2(F_1 - F_2)}$$

where   $D$ is the distance (in feet) from measurement end to open or short,
        $K$ is the velocity constant in percentage of the free-space propaga-
           tion velocity of radio waves (Table 11-1),
        $F_1$ is frequency of the first dip (in megaherz), and
        $F_2$ is frequency of the second (adjacent) dip (in megahertz).

### NOTE

It should be noted that the distance calculated is from the coupling
loop end to the first open or short in the line. For example, if a 100-
foot line is measured and this line has a short at 70 feet, the distance
measured will be 70 feet, not 100 feet.

## 11-5. Standing-Wave Ratio Measurements

The standing-wave ratio (SWR) of an antenna is actually a measure
of match or mismatch between the antenna and the transmission line.
When an antenna and the transmission line are perfectly matched as to
impedance, all of the energy or signal will be transferred to or from the
antenna and there will be no loss. If here is a mismatch (as is the case
in any practical application), some of the energy will be reflected back into
the line. This energy will cancel part of the desired signal. If the voltage
(or current) is measured along the line, there will be voltage or current
maximums (where the reflected signals are in phase with the outgoing
signals) and voltage or current minimums (where the reflected signal is
out of phase, partially canceling the outgoing signal). These maximums
and minimums are called *standing waves*. The ratio of the maximum to
the minimum is the standing-wave ratio, or SWR. The ratio can be related to either voltage or current. Since voltage is easier to measure it is
usually used, resulting in the common term *voltage standing-wave ratio,*
or VSWR. The calculations for VSWR are shown in Fig. 11-13.

A standing-wave ratio of 1-to-1 means that there are no maximums or
minimums (the voltage is constant at any point along the line) and that
there is a perfect match between antenna and transmission line.

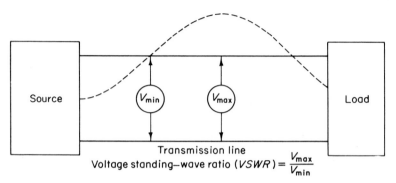

Fig. 11-13. Calculations for voltage standing-wave ratio (VSWR).

As discussed in Chapter 10, the SWR of high-frequency (microwave) signals can be found by physical measurement of the waves, using slotted lines, Lecher lines, etc. In the case of microwave frequencies, a meter is physically moved along the line to measure maximum and minimum voltages (or currents). This is not practical at lower frequencies due to the physical length of the waves. At lower frequencies it is more practical to measure forward or outgoing voltage and reflected voltage and then calculate the *reflection coefficient* (reflected voltage/outgoing voltage). The reflection coefficient can be converted to SWR if desired.

There are a number of SWR meters used in communications or antenna work. The combination SWR/power meters found in amateur radio and citizen's band use are typical. Most of these meters use a form of directional coupler to measure forward voltage and reflected voltage. Some of the meters read out in SWR; others read out in reflection coefficient. Still others read out in voltage (forward and reflected) that can be converted to SWR or reflection coefficient.

Basic SWR meter circuits are quite simple, and it is possible to fabricate them in the laboratory or shop. However, it is not practical in most cases to do so. The basic circuit requires that a directional coupler and pickup or probe be inserted in the transmission line. Even under good conditions, a mismatch and some power loss may result. A poorly designed and fabricated pickup can result in considerable power loss, as well as inaccurate indications. Therefore, it is better to use commercial SWR meters (as it is better to use commercial bridge circuits) for practical considerations.

The basic SWR/meter circuit is shown in Fig. 11-14. Operation of the circuit is as follows. As shown, there are two pickup wires, both parallel to the center conductor of the transmission line. Any RF voltage on either of the parallel pickups is rectified and applied to the meter through the selector switch $S_1$. Each pickup wire is terminated in the im-

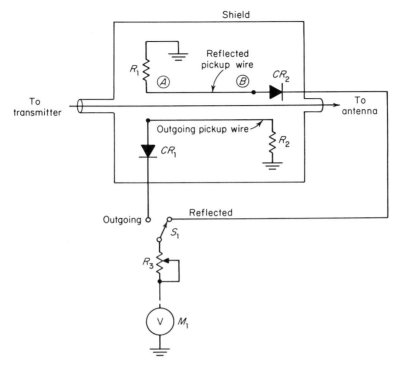

**Fig. 11-14.** Typical SWR meter circuit used in communication antenna measurements.

pedance of the transmission line by corresponding resistors $R_1$ and $R_2$.

The outgoing voltage (transmitter to antenna) is absorbed by $R_1$. Therefore, there is no outgoing voltage on the reflected voltage pickup wire beyond point $A$. However, the outgoing voltage remains at the transmission line and at the outgoing voltage pickup. This voltage is rectified by $CR_1$ and appears as a reading on meter $M_1$ when $S_1$ is in the outgoing voltage position.

The opposite condition occurs for the reflected voltage (antenna to transmitter). There is no reflected voltage on the outgoing pickup wire beyond point $B$ because the reflected voltage is absorbed by $R_2$. The reflected voltage does appear on the reflected pickup wire beyond this point; it is rectified by $CR_2$ and appears as a reading on meter $M_1$ (with $S_1$ in the reflected voltage position).

Once the outgoing and reflected voltages have been found, the reflection coefficient (not SWR) is determined using the equations of Section 10-4.2. If desired, the reflection coefficient can be converted to SWR using equations of the same Section 10-4.2.

In many commercial SWR meters, the meter scale is calibrated in SWR, although the actual readout is in reflection coefficient. With such systems switch $S_1$ is set to read outgoing voltage, and resistor $R_3$ is adjusted until the meter needle is aligned with some "set" or "calibrate" line (near the right-hand end of the meter scale). Switch $S_1$ is then set to read reflected voltage, and the meter needle moves to the corresponding SWR indication.

# CHAPTER 12

# Miscellaneous Circuit and
# Component Tests

## 12-1. Photocell Tests

There are two basic types of photocells: *photovoltaic* and *photoconductive*. The photovoltaic cells produce an output voltage and current in the presence of light and are often called *solar batteries* or solar cells. The photoconductive cells are often termed *light-sensitive resistors,* since they function as a resistor and do not generate an output. Instead, photoconductive cells act as a resistance that varies in the presence of light, thus changing the amount of current being conducted through the circuit.

The basic test circuit for both photovoltaic and photoconductive cells is shown in Fig. 12-1.

In either circuit, the cell is exposed to sunlight or artificial light, and the meter reading is noted. Usually, a single cell will produce sufficient output for a readable indication on the low-voltage range of a VOM or electronic voltmeter. However, some cells may require the output to be read as a current. For example, some cells produce less than 1 mA when exposed to strong sunlight. Always check the manufacturer's data for cell characteristics.

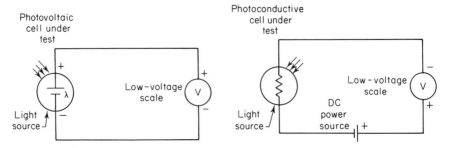

**Fig. 12-1.** Basic photocell test circuits.

## 12-2. Relay Tests

An oscilloscope can be used to check the make and break of relay contacts. The presence of contact "bounce" as well as actual make-time and break-time of the contacts can be displayed and measured. To be effective, the oscilloscope should be capable of *single sweep operation*. Also, because of the instantaneous nature of the trace, the display should be photographed (unless a *storage-type* oscilloscope is used).

Figure 12-2 shows the basic test connection diagrams for both a-c and d-c relays. The procedure is as follows.

1. Connect the equipment as shown in Fig. 12-2a or 12-2b, as applicable.

2. Place the oscilloscope in operation as described in the instruction manual. Set the oscilloscope to single sweep mode. Set the sync selector to external or as necessary so that the oscilloscope will be triggered by the external d-c voltage. Set up the oscilloscope camera as necessary.

3. If the trace is to be photographed, hold the camera shutter open, close and open switch $S_1$, then close the camera shutter and develop the picture.

4. Using the developed photo, measure the bounce (if any) amplitude along the vertical axis and the bounce duration along the horizontal axis.

5. Measure the make-time (interval between application of voltage and actual closure of contacts) and break-time (interval between interruption of voltage and opening of the relay contacts) along the horizontal axis.

## 12-3. Vibrator Tests

An oscilloscope can be used to check the make and break of vibrator contacts as well as to display the operation of synchronous and non-synchronous vibrator power supplies under actual operating conditions.

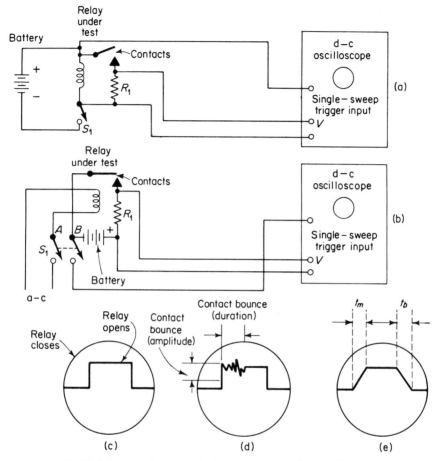

**Fig. 12-2.** Testing relays for make-time, break-time, and contact bounce.

Vibrators are best tested by observing contact operation with the vibrator in the power supply. The approximate square wave across the power supply transformer primary is displayed on the oscilloscope and is compared against an "ideal" waveform. The procedure is as follows.

1. Connect the equipment as shown in Fig. 12-3.
2. Place the oscilloscope in operation as described in the instruction manual. Switch on the internal recurrent sweep. Set the sync selector and sweep selector to internal. Adjust the sweep frequency and sync controls for one (or preferably two) stationary cycles on the oscilloscope screen.
3. Compare the actual oscilloscope display with the "ideal" display of

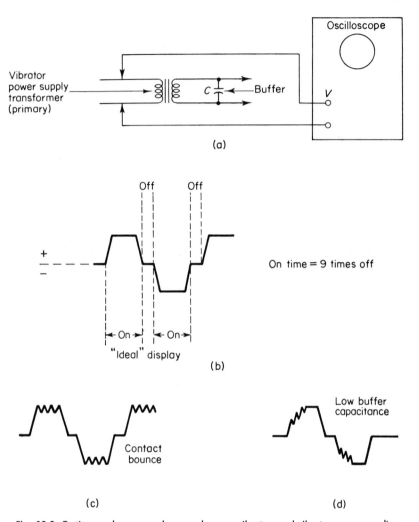

**Fig. 12-3.** Testing synchronous and nonsynchronous vibrators and vibrator power supplies.

Fig. 12-3b. Measure the on-time and off-time of the display along the horizontal axis.

4. Usually, the display amplitudes (+ and —) should be equal, although the actual amplitude is not critical.

5. In most vibrator power supplies, the on-time total should be approximately nine times the off-time total. (The on-time intervals represent the length of time the vibrator contacts are actually closed and delivering current. The off-time intervals represent the length of time the vibrator contacts are open.)

6. In addition to showing the power supply efficiency (on-time to off-time percentage), the waveform display can also show such conditions as contact bounce (Fig. 12-3c), or insufficient buffer capacitance (Fig. 12-3d).

## 12-4. Chopper Tests

An oscilloscope can be used to check the operation of electromechanical and electronic choppers. Such choppers are similar in function to vibrators in that they convert dc into ac for further processing (usually amplification). A typical example of chopper use is to convert d-c signals into ac for application to an a-c amplifier. Electromechanical choppers differ from vibrators in that they are driven by an external source independent of the d-c source to be converted. The resultant output is a square wave that is proportional to the d-c input. Electronic choppers are essentially electronic switches that produce a square-wave output proportional to d-c input. Either way, an oscilloscope provides a quick test of operation. The procedure is as follows.

1. Connect the equipment as shown in Fig. 12-4.

2. Place the oscilloscope in operation as described in the instruction manual. Switch on the internal recurrent sweep. Set the sync selector and sweep selector to internal. Adjust the sweep frequency and sync controls for one (or preferably two) stationary cycles on the oscilloscope screen.

3. Measure the display amplitude along the vertical axis. Measure the on-time and off-time of the display along the horizontal axis.

4. It is also possible to check the chopper output for electrical noise. All test connections remain the same, except that the d-c input voltage is removed. Under these conditions, any vertical deflection is the result of electrical noise.

### NOTE

An electronic chopper is tested for noise with the d-c input terminals open. An electromechanical chopper should be tested both ways, first with the d-c input terminals shorted, then open.

## 12-5. Magnetic Component Tests

Hysteresis and saturation are properties of magnetic components that can be tested using an oscilloscope. Both the vertical and horizontal channels of the oscilloscope must be voltage calibrated. Also, the horizontal

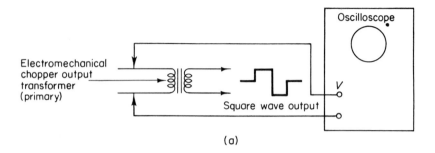

(a)

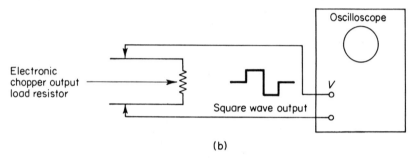

(b)

**Fig. 12-4.** Testing electromechanical and electronic choppers.

and vertical channels must be identical, or nearly identical, to eliminate any phase difference. The procedure is as follows.

1. Connect the equipment as shown in Fig. 12-5.

2. Place the oscilloscope in operation as described in the instruction manual. Voltage-calibrate both the vertical and horizontal channels as necessary. The oscilloscope trace (spot) should be at the vertical and horizontal center with no signal applied to either channel.

3. Switch off the internal recurrent sweep. Set sweep selector and sync selector to external. Leave the horizontal and vertical gain controls set at the (voltage) calibrate position.

4. Set the generator to the desired test frequency. Increase the generator output, noting that the pattern enlarges on the screen both horizontally and vertically. Continue to increase generator output until the upper and lower ends of the pattern bend as saturation is reached.

5. Using the voltage-calibrated vertical axis, measure the peak voltage at which saturation occurs. This voltage can be converted to current (through the device under test), using the value of $R_1$ in a basic Ohm's law equation. For example, if the vertical deflection voltage is 10 mV and the

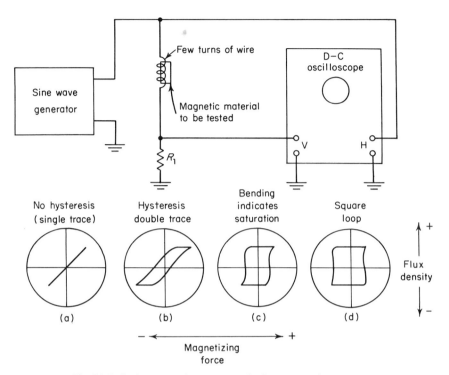

**Fig. 12-5.** Testing magnetic components for hysteresis and saturation.

value of $R_1$ is 1 ohm, the current through the magnetic component under test is 10 mA.

6. Using the voltage-calibrated horizontal axis, measure the peak voltage at which saturation occurs.

7. Compare the voltage and current values of the hysteresis loop against the manufacturer's specifications. Not all magnetic materials are supposed to show hysteresis. On those devices for which hysteresis is normal, the square loop (Fig. 12-5d) is usually considered as ideal. The loop area can be readily measured using the oscilloscope screen divisions.

## 12-6. Ferroelectric Component Tests

Many ferroelectric materials exhibit electrostatic hysteresis and saturation. It is possible to measure the hysteresis and saturation of these ferroelectric materials using an oscillosocpe. The procedure is almost identical to that used to test magnetic components as described in Section 12-5. The major differences in the procedure are that the equipment is connected as shown in Fig. 12-6, a fixed capacitor $C_1$ is used in place of resistor $R_1$, and

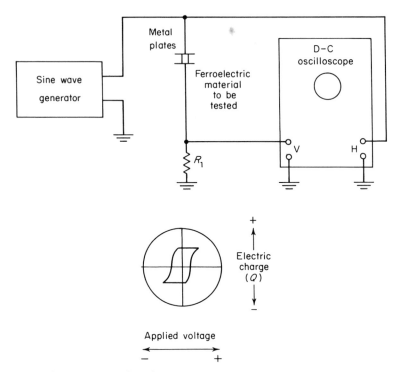

**Fig. 12-6.** Testing ferroelectric components for hysteresis and saturation.

vertical deflection is proportional to the voltage developed across $C_1$. If the ferroelectric component is a material rather than a complete piece of equipment, a flat slab of the material can be placed between two metal plates.

Vertical deflection is proportional to electric charge. Horizontal deflection is proportional to applied voltage.

## 12-7. Microphone Tests

A microphone can be checked for distortion and frequency response using an oscilloscope. Distortion is checked by applying a sine wave to a loudspeaker placed near the microphone and monitoring the microphone output on an oscilloscope. Frequency response is checked in essentially the same way, by varying the sine wave over the desired test range of the microphone.

1. Connect the equipment as shown in Fig. 12-7.
2. Place the oscilloscope in operation as described in the instruction

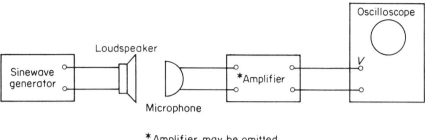

**Fig. 12-7.** Testing microphones with an oscilloscope.

manual. Switch on the internal recurrent sweep. Set the sweep selector and sync selector to internal.

3. Place the generator and amplifier (if any) in operation as described in their instruction manuals. Set the generator output frequency to the low limits of the microphone. Adjust the generator and amplifier controls for a suitable pattern on the oscilloscope.

4. Check the oscilloscope pattern for distortion. If there is doubt as to the origin of any observed distortion, temporarily connect the generator output to the amplifier input. If distortion is removed, the cause is in the microphone (assuming that the loudspeaker is distortion free).

#### NOTE

For the most accurate results, the microphone should be shielded from all sound sources except the loudspeaker.

5. Without changing the generator output amplitude, vary the generator frequency over the entire test range of the microphone. Check for any change in amplitude on the oscilloscope pattern.

#### NOTE

A response curve can be made for the microphone by following the procedures of Section 8-12.

### 12-8. Transformer Tests

A VOM or electronic voltmeter is the ideal instrument to check a transformer. The obvious test is to measure the transformer windings for opens, shorts, and the proper resistance value with an ohmmeter. If the ohmmeter is equipped with a high-ohms adapter (Section 3-6), it is possible to check a transformer for leakage between windings. In addition to basic resistance checks, it is possible to test a transformer's proper polarity markings, regulation, impedance ratio, and center-tap balance with a voltmeter. If the transformer is tuned (such as an IF transformer), the Q factor can be checked with a meter. However, since a tuned transformer can be considered as a circuit, the procedure for measuring Q is described in later sections of Section 12-12.

### 12-8.1. Checking Transformer Phase Relationships

When two supposedly identical transformers must be operated in parallel and the transformers are not marked as to phase or polarity, the phase relationship of the transformers can be checked using a voltmeter and a power source. The test circuit is shown in Fig. 12-8. For power transformers, the source should be line voltage (115 V). Other transformers can be tested with lower voltage dropped from a line source or from an audio generator.

The transformers are connected in proper phase relationship if the meter reading is zero or very low. The transformers are out-of-phase if the secondary output voltage is double that of the normal secondary output. This condition can be corrected by reversing either the primary or secondary leads (but not both) of one transformer (but not both transformers).

If the meter indicates some secondary voltage, it is possible that one transformer has a greater output than the other. This condition will result in considerable local current flowing in the secondary winding and will produce a power loss (if not actual damage to the transformers).

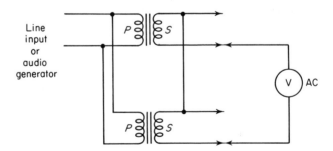

**Fig. 12-8.** Testing transformer phase relationships.

### 12-8.2. Checking Transformer Polarity Markings

Many transformers are marked as to polarity or phase. These markings may consist of dots, color-coded wires, or some similar system. Unfortunately, transformer polarity markings are not always standard. Generally, transformer polarities are indicated on schematics as dots next to the terminals. When standards are used, the dots mean that if electrons are flowing *into* the terminal with the dot, the electrons will flow *out of* the secondary terminal with the dot. Therefore, the dots have the same polarity so far as the external circuits are concerned. No matter what system is used, the dots or other markings show *relative phase,* since instantaneous polarities are changing across the transformer windings.

From a practical standpoint, there are only two problems of concern: the relationship of the primary to the secondary and the relationship of markings on one transformer to those on another.

The phase relationship of primary to secondary can be determined using the test circuit of Fig. 12-9. First check the voltage across terminals 1 and 3 and then across 1 and 4 (or 1 and 2). Assume that there are 3 V across the primary, with 7 V across the secondary. If the windings are as shown in Fig. 12-9a, the 3 V will be added to the 7 V and will appear as 10 V across terminals 1 and 3. If the windings are as shown in Fig. 12-9b, the voltages will oppose and will appear as 4 V (7 − 3) across terminals 1 and 3.

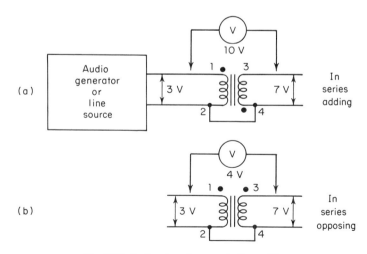

**Fig. 12-9.** Testing transformer polarity markings.

The phase relationship of one transformer marking to another can be determined using the test circuit of Fig. 12-10. Assume that there is a 3-V output from the secondary of transformer $A$ and a 7-V output from transformer $B$. If the markings are consistent on both transformers, the two voltages will oppose, and 4 V will be indicated. If the markings are *not* consistent the two voltages will add, resulting in a 10-V reading.

### 12-8.3. Checking Transformer Regulation

All transformers have some regulating effect. That is, the output voltage of a transformer tends to remain constant with changes in load. Regulation is usually expressed as a percentage and is equal to

$$\text{percentage of regulation} = \frac{\text{no-load voltage} - \text{load voltage}}{\text{no-load voltage}}$$

Some transformers are designed to provide good regulation (a low percentage). Other transformers show very poor regulation (a high percentage).

Transformer regulation can be tested using the circuit of Fig. 12-11. The value of $R_1$ (load) should be selected to draw the maximum rated current from the secondary.

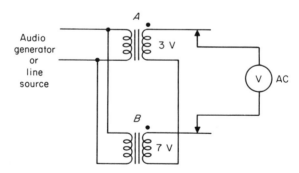

Fig. 12-10. Testing consistency of transformer polarity or phase markings.

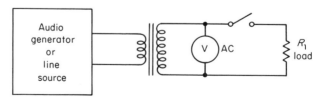

Fig. 12-11. Testing transformer regulation.

First, measure the secondary output voltage without a load, and then with a load. Use the equation to find percentage of regulation.

### 12-8.4. Checking Transformer Impedance Ratio

The impedance ratio of a transformer is the square of the winding ratio. Impedance ratio should not be confused with impedance of a transformer. Impedance measurements are discussed in Chapter 6.

If the winding ratio of a transformer is 15-to-1, the impedance ratio is 225-to-1. Any impedance value placed across one winding will be reflected onto the other winding by a value equal to the impedance ratio. For example, assume an impedance ratio of 225-to-1 and an 1800-ohm impedance placed on the primary. The secondary would then have reflected impedance of 8 ohms. Likewise, if a 10-ohm impedance were placed on the secondary, the primary would have a reflected impedance of 2250 ohms.

Impedance ratio is related directly to turns ratio (primary to secondary). However, turns ratio data are not always available, so the ratio must be calculated using a test circuit as shown in Fig. 12-12.

Use line voltage as a source for power transformers and an audio generator for other transformers.

Measure both the primary and secondary voltage. Divide the larger voltage by the smaller, noting which is primary and which is secondary. For convenience set either the primary or secondary to some exact voltage.

The *turns ratio* is equal to one voltage divided by the other.

The *impedance ratio* is the square of the turns ratio.

For example, assume that the primary shows 115 V, with 23 V at the secondary. This indicates a 5-to-1 turns ratio and a 25-to-1 impedance ratio.

### 12-8.5. Checking Transformer Winding Balance

There is always some imbalance in center-tapped transformers. That is, the turns ratio and impedance are not exactly the same on both sides of the center tap. The imbalance is usually of no great concern in shop-type

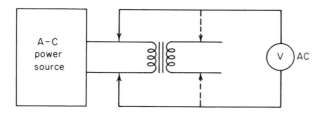

**Fig. 12-12.** Measuring transformer impedance ratio.

equipment but can be critical in laboratory-type transformers. It is possible to find a large imbalance by measuring the d-c resistance on either side of the center tap. However, a small imbalance might not show up, especially if the d-c resistance is high.

It is usually more practical to measure the *voltage* on both sides of a center tap, as shown in Fig. 12-13. If the voltages are equal, the transformer winding is balanced. If a large imbalance is indicated by a large voltage difference, the winding should then be checked with an ohmmeter for shorted turns or some similar failure.

## 12-9. Battery Tests

The obvious test for a battery is to measure the voltage from all of the cells or from each cell on an individual basis. Such a test will not show how a battery will maintain its voltage output under load. Therefore, it is necessary to test a battery under dynamic conditions. The following paragraphs describe these procedures.

### 12-9.1. Measuring Battery Output Under Load

1. Connect the battery, load, and meter as shown in Fig. 12-14. If the battery is to be tested out-of-circuit, use a load resistance that will produce the maximum-rated current flow.

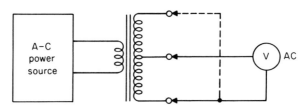

Fig. 12-13. Testing transformer winding balance.

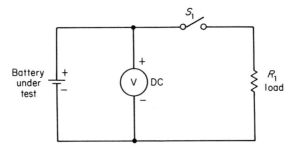

Fig. 12-14. Measuring battery output under load.

2. Measure the battery voltage both with and without a load. Note any drop in voltage when the load is applied. If the battery is to be tested in-circuit, measure the battery voltage with the equipment turned off. Then turn the equipment on, and measure the full-load voltage.

3. Normally, there will be not more than a 10% or 15% drop in voltage output when a full load is applied. The exact amount of voltage drop will depend upon the type of battery. Also, the output voltage of a good battery will return to full value when the load is removed. A defective battery (or cell) will drop in output voltage when a load is applied and will remain low.

4. A typical lead-acid storage battery will produce an output of approximately 2.1 V without a load, and 1.75 V under full load. The condition of a storage battery can also be checked using a hydrometer (to measure specific gravity).

5. Always use the lowest practical voltage scale to measure battery voltage. This is necessary since a 0.1-V difference can be important in the single cell of a battery.

### 12-9.1. Locating a Defective Battery Cell

When it is necessary to locate a suspected cell in a group of many identical cells, connect all of the cells in series across a load. (Fig. 12-15.) Remove the load. Then measure the voltage across each cell. The defective cell will show a lower output than the remaining cells or will possibly show zero output. In some cases the polarity of the defective cell may reverse.

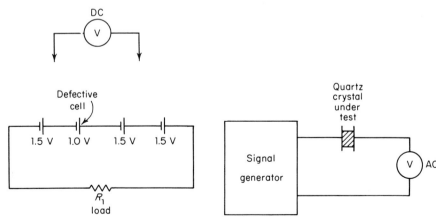

**Fig. 12-15.** Locating a defective battery cell.

**Fig. 12-16.** Measuring approximate resonant frequency of quartz crystal.

## 12-10. Quartz Crystal Tests

The *approximate* resonant frequency of a quartz crystal can be found using the test circuit of Fig. 12-16.

1. Set the meter to measure a-c voltage.
2. Adjust the signal generator output to the supposed frequency of the crystal. Then adjust the signal generator frequency for maximum indication on the meter. Read the crystal frequency from the signal generator controls (frequency dial or counter). Be careful not to increase the signal generator output voltage beyond the maximum rated limits of the crystal. Excess voltage can crack or otherwise damage the crystal.

## 12-11. L and T Pad Tests

$L$ and $T$ pads, whether considered as a circuit or a component, require special measurement procedures. Generally, pads can be checked using an ohmmeter, but if it is suspected that a pad is producing some reactance (inductive or capacitive), the pad must be checked with an audio source.

### 12-11.1. Measuring L Pads for Impedance

Usually the input impedance of an $L$ pad will remain constant, but the output impedance will vary as the pad setting is changed (or the pad can be reversed so that the output impedance is constant with the input impedance varying).

The input impedance test circuit is shown in Fig. 12-17. The load resistance $R_L$ should be equal to the normal load as seen from the output side of the pad. Vary the pad throughout its entire range and note the resistance values found at the input circuit. Assuming no reactance, the d-c input resistance values will equal the input impedance values.

The output impedance test circuit is shown in Fig. 12-18. The load

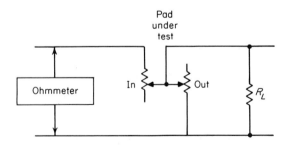

**Fig. 12-17.** Measuring L pad input impedance.

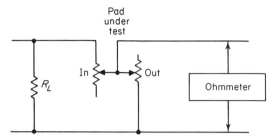

**Fig. 12-18.** Measuring L pad output impedance.

resistance $R_L$ should be equal to the normal load as seen from the input side of the pad. Vary the pad throughout its entire range and note the resistance values obtained at the output circuit. Assuming no reactance, the d-c output resistance values will equal the output impedance values.

### 12-11.2. Measuring T Pads for Impedance

The input and output impedance of a $T$ pad should remain constant as the pad setting is changed (within a specified tolerance). This is one of the advantages of a $T$ pad over an $L$ pad. Note, however, that the input and output impedances of a $T$ pad are not necessarily the same. Therefore, a $T$ pad can also be used to match impedances as well as to vary signal strength.

The impedance test circuit is shown in Fig. 12-19. In Fig. 12-19a (to measure input impedance) the load resistance $R_L$ (equal to the normal load as seen from the output side of the pad) is connected at the output with the ohmmeter at the input. The circuit is reversed in Fig. 12-19b (to measure output impedance).

In either circuit, vary the pad throughout its entire range and note that the resistance remains constant (within tolerance). Assuming no reactance, the resistance values will equal the impedance values.

### 12-11.3. Measuring Pads for Reactive Effect

An $L$ or $T$ pad can be checked for reactive effect using the circuit of Fig. 12-20. The load resistance $R_L$ should be equal to the rated output impedance of the pad (or the impedance that will be seen from the output side). The input resistance $R_{in}$ should be equal to the rated input impedance of the pad. However, if the pad impedance is close to that of the audio generator impedance, then $R_{in}$ should be added to the generator impedance to equal the pad input impedance. For example, if the pad input impedance is 10 k$\Omega$ and the audio generator output impedance is 50 ohms,

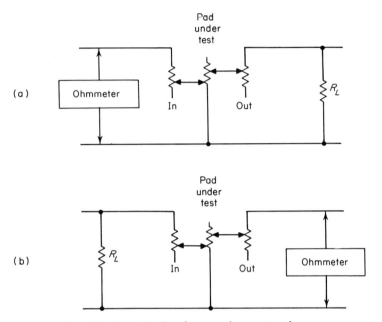

(a)

(b)

**Fig. 12-19.** Measuring *T* pad input and output impedance.

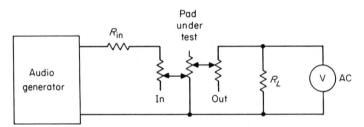

**Fig. 12-20.** Testing pads for reactive effect.

$R_{in}$ should be approximately 10 k$\Omega$. If the pad input is 100 ohms with a 50-ohm generator output, $R_{in}$ should be 50 ohms.

    1. Adjust the audio generator output voltage for a good reading on the meter.

    2. Vary the audio generator output over the entire frequency range in which the pad will be used. *Do not* change the generator output *voltage.*

    3. If the pad has any reactive effect the meter reading will not be constant but will rise (or fall) as some particular frequency or over some particular frequency range.

    4. Repeat the test at various settings of the pad.

## NOTE

When making this test or any test that involves varying the output of a generator, remember that all generators do not have a flat output (constant voltage output over the entire frequency range). The same is true of meters. This can lead to an error in judgment. The ideal remedy is to calibrate the generator over its entire frequency range and record any variations. When this is not practical, a quick check can be made by connecting the meter directly to the generator output, varying the generator over the frequency range, and then noting any variation in output voltage. These variations can then be compared with any variations found with the meter connected at the pad output. If the variations are the same, it is likely that the generator or meter is at fault, not the pad.

## 12-12. Resonance and Q Measurements

When a capacitance and an inductance are connected in series or parallel and a signal of appropriate frequency is applied, the circuit will be *resonant*. The inductive reactance will be equal to the capacitive reactance. In a series circuit, the impedance will be zero, theoretically (in practice there will be some pure resistance that will act as an impedance). In a parallel circuit, the impedance will be infinite.

Resonance is dependent upon three variables: capacitance, inductance, and frequency. The calculations for resonance are given in Fig. 12-21. The practical procedures for measurement of resonance and resonant frequency are discussed in Chapter 8.

A resonant circuit has a Q, or *quality factor,* as do individual capacitors and inductors.

*The Q of an inductor* (coil) is determined by the ratio of inductive reactance to resistance

$$Q = \frac{X_L}{R}$$

The greater the inductive reactance to a given resistance, the greater the Q.

*The Q of a capacitor* is determined by the ratio of capacitive reactance to resistance

$$Q = \frac{X_C}{R}$$

The greater the inductive reactance to a given resistance, the greater the Q.

*The Q of a resonant circuit* is dependent upon the individual Q factors of the inductance and capacitance used in the circuit. This applies to a tuned component (such as a tuned transformer) as well as a complete circuit (such as a tuned RF or IF stage of a receiver).

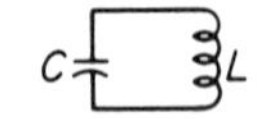

$$F = \frac{1}{6.28 \sqrt{LC}}$$

Parallel*
(infinate impedance)

Series
(zero impedance)

Frequency $(F)$ in kilohertz (KHz) $= \dfrac{10^6}{6.28 \sqrt{L} \text{ (in microhenrys)} \times C \text{ (in picofarads)}}$

or

Frequency $(F)$ in kilohertz (KHz) $= \dfrac{159.2}{\sqrt{L} \text{ (in microhenrys)} \times C \text{ (in microfarads)}}$

$L$ (in microhenrys) $= \dfrac{25,330}{\text{Frequency (in KHz)}^2 \times C \text{ (in microfarads)}}$

$C$ (in microfarads) $= \dfrac{25,330}{\text{Frequency (in KHz)}^2 \times L \text{ (in microhenrys)}}$

* For parallel resonant circuits, the
equations are a close approximation
and should be accurate when circuit
$Q$ is 10 or higher.

**Fig. 12-21. Calculations for resonant circuits.**

If a circuit had pure inductance and capacitance, the response curve would be very sharp (or high Q). However, since any resonant circuit has some resistance, the Q factor is limited. A high Q is not always desired.

The primary concern in a resonant circuit or stage is that a high Q will produce a sharp resonant curve, whereas a low Q will produce a broad response curve.

Usually a resonant circuit is measured at points on either side of the resonant frequency where the signal amplitude is down to 0.707 of the peak resonant value. Other reference points can be used in special cases.

### 12-12.1. Practical Q Measurement

The Q of a circuit can be measured using a signal generator and a meter with an RF probe. An electronic voltmeter will provide the least loading effect on the circuit and will therefore provide the most accurate indication.

Figure 12-22a shows the test circuit in which the signal generator is connected directly to the input of a complete stage, and Fig. 12-22b shows the indirect method of connecting the signal generator to the input.

When the stage or circuit has sufficient gain to provide a good reading on the meter with a nominal output from the generator, the indirect method is preferred. Any signal generator will have some output impedance (such

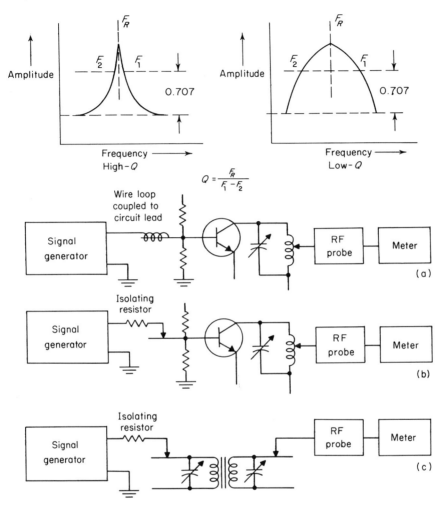

$$Q = \frac{F_R}{F_1 - F_2}$$

Fig. 12-22. Measuring circuit and component Q.

as a 50-ohm output resistor). When this resistance is connected directly to the tuned circuit, the Q is lowered, and the response becomes broader. (In some cases, the generator output impedance will seriously detune the circuit.)

Where it is not practical to use the indirect method, the generator output should be connected through an isolating resistor $R$. The value of $R$ is not critical but should be near that of the stage ahead of the circuit being measured.

Figure 12-22c shows the test circuit for a single component (such as an IF transformer). Since there is rarely enough gain in a transformer, the direct method (with an isolating resistor) must be used.

1. Connect the equipment as shown in Fig. 12-22a, b, or c, as applicable.

2. Tune the signal generator to the circuit resonant frequency. Operate the generator to produce an unmodulated output.

3. Tune the signal generator frequency for maximum reading on the meter. Note the generator frequency.

4. Tune the signal generator below resonance until the meter reading is 0.707 of the maximum reading. Note the generator frequency.

### NOTE

To make the calculation more convenient, adjust the signal generator output so that the meter reading is some even value such as 1 V or 10 V after the generator is tuned for maximum. This will make it easy to find the 0.707 mark.

5. Tune the generator above resonance until the meter reading is 0.707 of the maximum reading. Note the generator frequency.

6. Calculate the circuit Q using the equation shown in Fig. 12-22.

7. For example, assume that the maximum meter indication occurred at 455 kHz ($F_R$), the below-resonance indication was at 453 kHz ($F_2$), and the above-resonance indication was at 458 kHz ($F_1$).

$$458 - 453 = 5, \qquad \frac{455}{5} = \text{a Q of 91}$$

## 12-13. Distortion Measurements

There are four basic methods of measuring distortion: sine-wave analysis, square-wave analysis, measurement of harmonic distortion, and measurement of intermodulation distortion.

In practice, analyzing sine waves to pinpoint distortion is a difficult job requiring considerable experience. Unless the distortion is severe, it may pass unnoticed. Therefore, sine waves are best used where harmonic distortion or intermodulation meters are combined with oscilloscopes for distortion analysis. If an oscilloscope is to be used alone, square waves provide the best basis for distortion analysis.

### 12-13.1. Practical Harmonic Distortion Analysis

No matter what amplifier circuit is used or how well the circuit is designed, there is always the possibility of odd or even harmonics being present with the fundamental. These harmonics combine with the fundamental and produce distortion, as is the case when any two signals are combined.

Commercial harmonic distortion meters operate on the *fundamental suppression* principle. As shown in Fig. 12-23, a sine wave is applied to the amplifier input, and the output is measured on the oscilloscope. The output is then applied through a filter that suppresses the fundamental frequency. Any output from the filter is then the result of harmonics. This output is also displayed on the oscilloscope, where the signal can be checked for frequency to determine the harmonic content. For example, if the input was 1 kHz and the output after filtering was 3 kHz, it would indicate third harmonic distortion.

The percentage of harmonic distortion can also be determined by this method. For example, if the output without filter was 100 mV and with filter was 3 mV, a 3% harmonic distortion would be indicated.

1. Connect the equipment as shown in Fig. 12-23.

**NOTE**

In some commercial harmonic distortion meters, the filter is tunable so that the amplifier can be tested over a wide range of fundamental

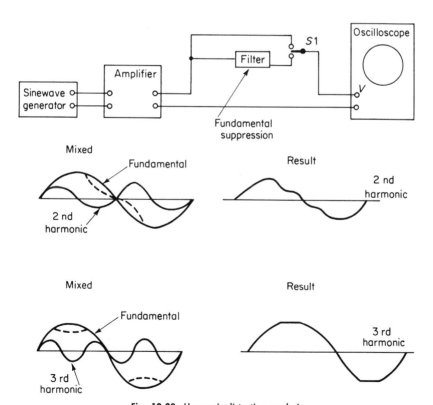

Fig. 12-23. Harmonic distortion analysis.

frequencies. In other harmonic distortion meters, the filter is fixed in frequency, but it can be detuned slightly to produce a sharp null.

2. Place the oscilloscope in operation as described in the instruction manual. Switch on the internal recurrent sweep. Set the sweep selector and sync selector to internal.

3. Set the amplifier amplitude and tone controls to their normal operating point or at the particular setting specified in the manufacturer's test data.

4. Place the generator in operation as described in the instruction manual. Set the generator output frequency to the filter null frequency. Set the generator output amplitude to the value recommended in the amplifier manufacturer's data. If specifications are not available, set switch $S_1$ to position 1 and increase the generator output until the waveform just starts to flatten, indicating that the amplifier is being overdriven. Then reduce the generator output until the waveform shows no distortion or flattening.

5. If necessary, adjust the sweep frequency controls to display a few cycles on the screen.

6. Measure the voltage with switch $S_1$ in position 1. Record this value as $E_1$.

7. Set switch $S_1$ to position 2. Adjust the filter for the deepest null indication on the oscilloscope. Record this value as $E_2$. (Note that some fundamental suppression filters are not tunable.)

8. Calculate the total harmonic distortion using the equation

$$D = 100 \frac{E_2}{E_1}$$

where  $D$ is percentage of total harmonic distortion,
    $E_1$ is output before filtering, and
    $E_2$ is output after filtering.

9. If the filter is tunable, select another frequency, tune the generator to that frequency, and repeat the procedure (Steps 4 to 8).

### 12-13.2. Practical Intermodulation Distortion Analysis

When two signals of different frequency are mixed in an amplifier there is a possibility of the lower-frequency signal amplitude-modulating the higher-frequency signal. This produces a form of distortion, known as *intermodulation distortion*.

Commercial intermodulation distortion meters consist of a signal generator and high-pass filter as shown in Fig. 12-24. The signal generator portion of the meter produces a high-frequency signal (usually about 7 kHz) that is modulated by a low-frequency signal (usually 60 Hz). The mixed signals are applied to the amplifier input. The amplifier output is connected through a high-pass filter to the oscilloscope vertical channel. The high-pass filter removes the low-frequency (60 Hz) signal. Therefore, the only signal appearing on the oscilloscope vertical channel should be the high-frequency (7 kHz) signal. If any 60 Hz signal is present on the display, it is being passed through as modulation on the 7-kHz signal.

Figure 12-24 also shows an intermodulation distortion test circuit that

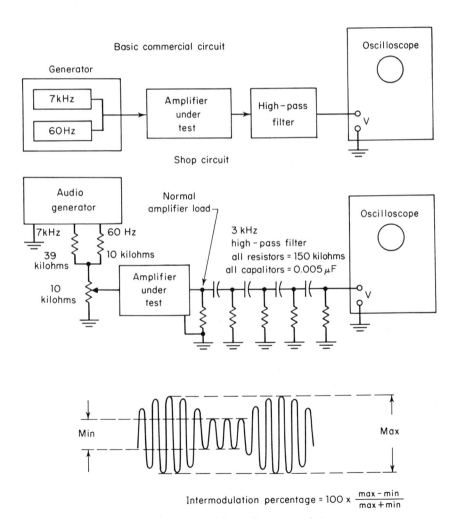

$$\text{Intermodulation percentage} = 100 \times \frac{\text{max} - \text{min}}{\text{max} + \text{min}}$$

**Fig. 12-24. Intermodulation distortion analysis.**

can be fabricated in the shop or laboratory. Note that the high-pass filter is designed to pass signals above approximately 3 kHz. The purpose of the 39 kΩ and 10 kΩ resistors is to set the 60-Hz signal at four times that of the 7-kHz signal. Most audio generators provide for a line-frequency output (60 Hz) that can be used as the low-frequency modulation source.

1. Connect the equipment as shown in Fig. 12-24.

2. Place the oscilloscope in operation as described in the instruction manual. Switch on the internal recurrent sweep. Set the sweep selector and sync selector to internal.

3. Set the amplifier amplitude and tone controls to their normal operating point or at the setting specified in the manufacturer's test data.

4. Place the generator in operation as described in the instruction manual.

5. If necessary, adjust the sweep frequency controls to display a few cycles on the screen.

6. If the laboratory circuit of Fig. 12-24 is used instead of a commercial meter, set the generator line-frequency output to 2 V (if adjustable). Then set the generator audio output (7 kHz) to 2 V. If the line-frequency output is not adjustable, measure the value and then set the generator audio output to the same value.

7. Calculate the total intermodulation distortion using the equation of Fig. 12-24.

8. If desired, repeat the intermodulation measurement at various settings of the amplifier gain and tone controls.

### 12-13.3. Practical Square-wave Distortion Analysis

The procedure for checking amplifier distortion by means of square waves requires some experience on the part of the operator. The primary concern is deviation of the amplifier (or stage) output waveform from the input waveform. If there is no change (except in amplitude), there is no distortion. If there is a change in the waveform, the nature of the change will often reveal the cause of distortion. That is where square waves are more effective than sine waves. Square waves have a high odd-harmonic content. Also, it is easier to see a deviation from a straight line with sharp corners than from a curving line.

As shown in Fig. 12-25, square waves are introduced into the amplifier input while the output is monitored on an oscilloscope. If the oscilloscope has a dual-trace feature, the input and output can be monitored simultaneously. The third, fifth, seventh, and ninth harmonics of a clean square

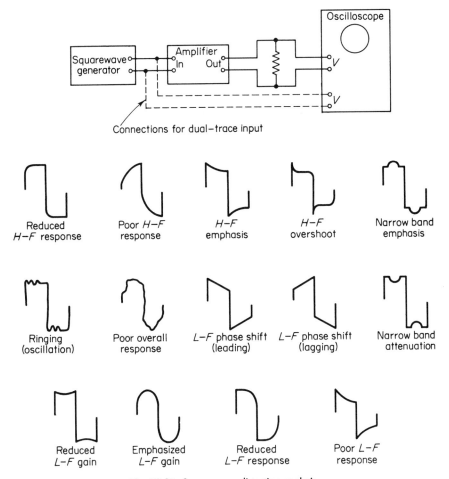

Fig. 12-25. Square-wave distortion analysis.

wave are emphasized. Therefore, if an amplifier passes a given audio frequency and produces a clean square-wave output, it is safe to assume that the frequency response is good up to at least nine times the fundamental frequency. For example, if an amplifier passes a clean square wave at 3 kHz, it shows a good response up to 27 kHz, which is beyond the top limit of the audio range.

1. Connect the equipment as shown in Fig. 12-25. A load resistor must be used for power amplifiers and should have a value equal to the amplifier's output impedance.

2. Place the oscilloscope in operation as described in the instruction

manual. Switch on the internal recurrent sweep. Set the sweep selector and sync selector to internal.

3. Set the amplifier amplitude and tone controls to their normal operating point or at the particular setting specified in the manufacturer's test data.

4. Place the generator in operation as described in the instruction manual. Set the generator output frequency to 1 kHz or as specified in the amplifier manufacturer's test data. If no specifications are available, increase the generator output until the waveform no longer increases in amplitude and/or shows distortion. Then reduce the generator output until the waveform shows no distortion.

5. If necessary, adjust the sweep frequency controls to display one (or possibly two) cycles on the screen.

6. Compare the amplifier output waveform with the input (generator output) waveform. If the output is identical to the input, except possibly for amplitude, there is no distortion. If the output is not identical, compare it with the typical response patterns of Fig. 12-25.

7. If desired, repeat the square-wave distortion analysis at other settings of the amplifier gain and tone controls, as well as other generator frequencies.

## 12-14. Noise Measurements

There are two basic types of noise measurement in electronics: electrical noise (which may include background noise, hum, transient noise, etc.) and acoustic noise (audible or inaudible sound).

### 12-14.1. Basic Electrical Noise Measurements

An oscilloscope provides a reliable means of measuring electrical "noise" or "hash." Such noise can be defined as any undesired signal present in a circuit. Usually, electrical noise is a combination of many frequencies and waveforms, all of which can be displayed on the oscilloscope. *Peak amplitude* of the noise is usually of major importance in any test. This can be measured with an oscilloscope on the voltage-calibrated vertical channel. In the case of transient noise, the display should be photographed (unless a storage-type oscilloscope is used).

As shown in Fig. 12-26, the noise source is connected to the oscilloscope vertical channel. The source can be connected directly to the vertical input or through a tuned amplifier. The tuned amplifier (such as is found in sound and vibration analyzers, wave analyzers, etc.) is used when noise frequency is of particular importance. If the noise source is a voltage, the vertical input can be connected across a component or branch of the circuit.

If the noise appears as a current, a resistance must be inserted in the circuit, and the voltage drop across the resistance measured. If it is not practical to interrupt the circuit, a current probe can be used.

1. Connect the equipment as shown in Fig. 12-26.

### NOTE

The tuned amplifier can be omitted if peak noise amplitude is the only factor of interest.

2. Place the oscilloscope in operation as described in the instruction manual. Set the sweep selector and sync selector to internal. Set up the oscilloscope camera as necessary.

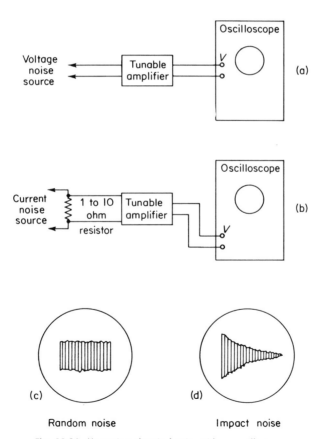

Random noise                    Impact noise

**Fig. 12-26.** Measuring electrical noise with an oscilloscope.

3. Set the sweep frequency and horizontal and vertical gain controls to display the noise pattern similar to that of Fig. 12-26.

4. Measure the peak-to-peak noise amplitude along the voltage-calibrated vertical axis.

5. If the noise pattern appears repetitive, measure the time interval between noise invervals, using the time-calibrated horizontal axis.

6. If *impact* noise is to be measured and recorded, hold the camera shutter open, initiate the impact, close the camera shutter, and develop the picture. Using the developed photo, measure the peak-to-peak amplitude as well as the time interval of the impact noise.

7. If it is desired to determine the frequency range of noise signals, tune the amplifier to each frequency of interest and note the time and amplitude of the noise signals.

### 12-14.2. Basic Background Noise Measurements

If the vertical channel of an oscilloscope is sufficiently sensitive, an oscilloscope can be used to check and measure the background noise level of an amplifier (or other circuit) as well as to check for the presence of hum, oscillation, etc. The oscilloscope vertical channel should be capable of a measurable deflection with about 1 mV (or less), since this is the background noise level of many amplifiers.

The basic procedure consists of measuring amplifier output with the gain control at maximum but without an input signal. The oscilloscope is superior to a voltmeter for noise level measurement since the frequency and nature of the noise (or other signal) are displayed visually.

1. Connect the equipment as shown in Fig. 12-27. The load resistor $R_1$ is used for power amplifiers and should have a value equal to the amplifier's output impedance.

2. Place the oscilloscope in operation as described in the instruction manual. Switch on the internal recurrent sweep. Set the sweep selector and sync selector to internal.

3. Set the amplifier gain control to maximum and the tone controls to their normal position, unless otherwise specified in the manufacturer's data.

4. Increase the oscilloscope vertical gain control until there is a noise or "hash" indication.

### NOTE

It is possible that a noise indication could be caused by pickup in the lead wires. If in doubt, disconnect the leads from the amplifier but not from the oscilloscope.

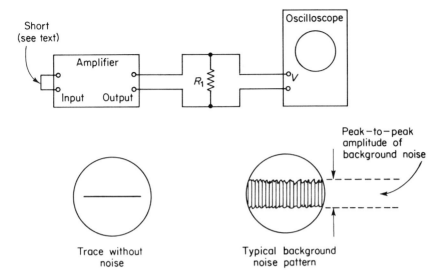

**Fig. 12-27.** Measuring amplifier noise and hum with an oscilloscope.

5. Measure the noise voltage. This is the total noise voltage, including hum, background noise, oscillation, etc.

6. If it is suspected that there is line hum present in the amplifier output, set the oscilloscope sync control to line. If a stationary signal pattern appears, it is due to line hum. Measure the amplitude of the line hum, if desired.

7. If a signal appears that is not at the line frequency it can be due to oscillation or stray pickup. Short the amplifier input terminals. If the signal remains, it is probably oscillation. In any case, the oscilloscope can be used to measure both the voltage and frequency of the unknown signal.

### 12-14.3. Basic Acoustic Noise Measurement

The procedure for measuring acoustic noise (or sound) with an oscilloscope is almost identical to that for electrical noise (Section 12-14.1). The major difference is that a microphone is used as the noise pickup and is not connected into a circuit to measure noise voltage or current. The microphone acts as a transducer and converts acoustic noise into an electrical signal. A capacitor-type microphone is used, with a preamplifier, for most acoustic noise measurements. Also, the voltage readings on the oscilloscope vertical channel can be converted into decibels (Chapter 7) if desired.

Set up the equipment as shown in Fig. 12-28 and follow the procedure of Section 12-14.1.

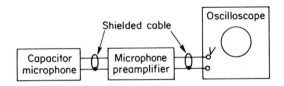

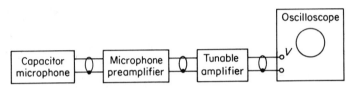

**Fig. 12-28.** Measuring acoustic noise (or sound) with an oscilloscope.

## 12-15. Modulation Measurements

An oscilloscope can be used to display the carrier of an amplitude-modulated radio wave. There are two basic methods: direct measurement of the modulation envelope and conversion of the envelope to a trapezoidal pattern. The trapezoidal method is the most effective, since it is easier to measure straight-line dimensions than curving dimensions. Also, non-linearity in modulation can be checked easily with the straight-line trapezoid. With either method, the percentage of modulation can be calculated from the dimensions of the modulation pattern.

### 12-15.1. Direct Measurement of the Modulation Envelope

1. Connect the equipment as shown in Fig. 12-29.

#### NOTE

If the vertical channel response is capable of handling the transmitter output frequency, the output can be applied through the oscilloscope vertical amplifier. If not, the transmitter output must be applied directly to the vertical deflection plates of the oscilloscope cathode-ray tube.

2. Place the oscilloscope in operation as described in the instruction manual. Switch on the internal recurrent sweep. Set the sweep selector and sync selector to internal. Adjust the horizontal and vertical gain controls for a no-signal trace.

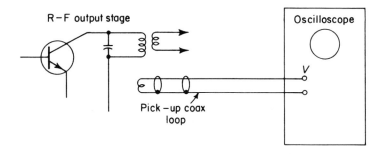

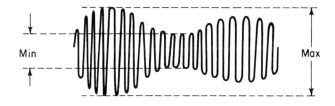

$$\text{Modulation (\%)} = 100 \times \left( \frac{max - min}{max + min} \right)$$

**Fig. 12-29.** Direct measurement of modulation envelope.

3. Place the transmitter in operation as described in the instruction manual. Initially, set the transmitter for an unmodulated carrier output.

4. Amplitude-modulate the transmitter with a sine-wave signal and check the oscilloscope pattern.

5. If necessary, adjust the sweep frequency controls to display a few cycles on the screen.

6. Measure the vertical dimensions MAX and MIN (Fig. 12-29) in screen divisions. Calculate the percentage of modulation using the equation of Fig. 12-29.

### 12-15.2. Trapezoidal Measurement of the Modulation Envelope

1. Connect the equipment as shown in Fig. 12-30.

### NOTE

Do not use the oscilloscope amplifiers. Make both the horizontal and vertical connections directly to the oscilloscope CRT.

2. Place the oscilloscope in operation as described in the instruction manual. Switch off the internal recurrent sweep. Set the sync selector to

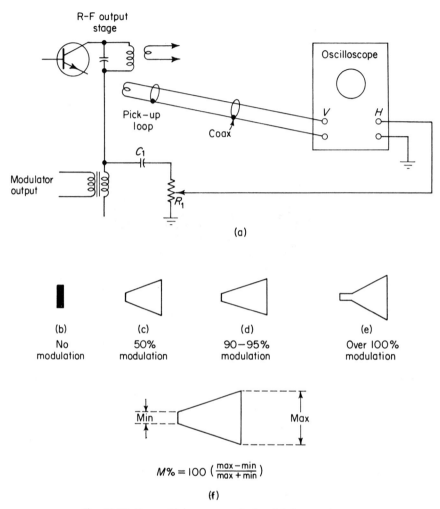

Fig. 12-30. Trapezoidal measurement of modulation envelope.

external. The no-signal trace should appear as a dot at the screen center.

3. Place the transmitter in operation as described in the instruction manual. Initially, set the transmitter for an unmodulated carrier output. This should produce a pattern as shown in Fig. 12-30b.

4. Amplitude-modulate the transmitter with a sine-wave signal and check the oscilloscope pattern. Figures 12-30c, d, and e show typical patterns for 50% modulation, 90 to 95% modulation, and overmodulation (over 100% modulation) respectively.

## NOTE

Adjust the oscilloscope display width with resistor $R_1$. The height of the oscilloscope display is adjusted by varying the coupling between the pickup coil and the output tank.

5. Observe the oscilloscope display for signs of nonlinearity. The straightness of the sides of the trapezoidal pattern indicate the modulation linearity. The trapezoidal pattern has the advantage that nonlinearity can be checked quickly.

6. Measure the vertical dimensions MAX and MIN (Fig. 12-30) in screen divisions. Calculate the percentage of modulation using the equation of Fig. 12-30.

## 12-16. Delay Measurements

The delay (or time interval) between two occurrences can be displayed and measured on an oscilloscope with a dual-trace feature. For example, the delay between an input pulse and an output pulse introduced by a delay line, digital circuit, multivibrator, or similar circuit, can be measured. Or there is the case in which two relays are supposed to close simultaneously but instead one or the other is delayed. Closure of the relay contacts will produce voltage pulses that can be applied to the oscilloscope vertical inputs, thereby providing a means of determining the extent of the delay.

If the delay is exceptionally short, the screen divisions can be calibrated with an external time-mark generator. If the oscilloscope has three vertical inputs, the timing wave from the time-mark generator can be displayed simultaneously with the input and output pulses.

1. Connect the equipment as shown in Fig. 12-31.

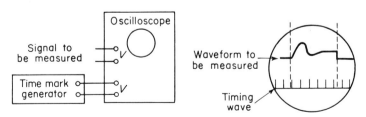

**Fig. 12-31.** Measuring delay with an oscilloscope.

2. Place the oscilloscope in operation as described in the instruction manual. Switch on the internal recurrent sweep. Set the sweep selector and sync selector to internal.

3. Switch on the delay device if it is powered.

4. Switch on the time-mark generator and pulse generator as described in their instruction manuals.

5. If the oscilloscope is a multiple-trace instrument, set the sweep frequency and sync controls for a single, stationary input and output pulse as shown in Fig. 12-31.

6. Set the horizontal and vertical gain controls and pulse generator output for desired pulse pattern width and height.

7. Count the timing spikes between the input and output pulses to determine the delay interval. Count the timing spikes between the beginning and end of each pulse to determine the pulse width or duration.

NOTE

If the oscilloscope is a dual-trace instrument, the screen divisions must be calibrated against a time-mark generator. Such generators produce pulse-type timing waves that are a series of sharp spikes spaced at precise time intervals. These spikes are applied to the vertical input and appear as a wavetrain as shown in Fig. 12-31. The oscilloscope horizontal gain and positioning controls are adjusted to align the timing spikes with screen lines until the screen divisions equal the timing pulses. The accuracy of the oscilloscope timing circuits is then of no concern, since the horizontal channel is calibrated against the external time-mark generator. The timing pulses can be removed and the signal to be measured applied to the vertical input, provided that the horizontal gain and positioning controls are not touched. Duration or time is read from the calibrated screen divisions in the normal manner. A typical time-mark generator will produce timing signals at intervals of 10, 1, and 0.1 $\mu$s.

## 12-17. Strain, Acceleration, and Pressure Measurements

The measurement of quantities such as strain, acceleration, and pressure is accomplished by means of a *transducer* and a readout. The readout can be a meter. However, an oscilloscope provides more reliable means of measuring *dynamic* quantities (where the quantities change rapidly with time). Because of the instantaneous nature of such measurements, the display should be photographed (unless a storage-type oscilloscope is used).

### 12-17.1. Strain Measurements

As shown in Fig. 12-32, strain gauges are connected in a bridge circuit, the output of which is applied to the oscilloscope vertical channel. The bridge is balanced under no-strain conditions with potentiometer $R_1$. One strain gauge (acting as a transducer) is placed on the material or structure to be tested; the other identical strain gauge is used as a reference. The oscilloscope is not deflected when the bridge is balanced (no-strain). When the material or structure is stressed, the resistance of the attached strain gauge is changed. This unbalances the bridge and produces a d-c output that is proportional to the change. The d-c output deflects the oscilloscope vertical trace and produces a plot of *strain versus time*. The bridge circuit can be calibrated in terms of strain (microinches of variation per inch-ounces of applied force), or strain versus voltage deflection, or whatever proves convenient for the particular test. The entire trace can be photographed for a permanent record.

1. Connect the equipment as shown in Fig. 12-32.

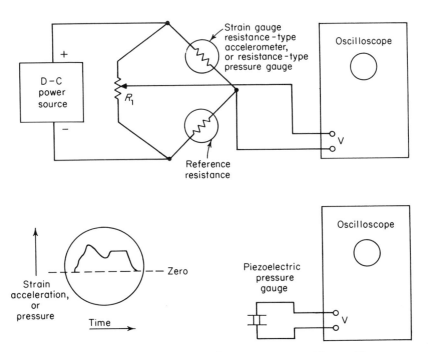

**Fig. 12-32.** Measuring dynamic strain, acceleration, or pressure with an oscilloscope.

2. Place the oscilloscope in operation as described in the instruction manual. Switch on the internal recurrent sweep. Set the sweep selector and sync selector to internal. Use a sweep time interval that will be longer than the strain interval. Set up the oscilloscope camera as necessary.

3. Set the oscilloscope to measure dc. Balance the bridge adjusting potentiometer $R_1$. The oscilloscope should be at zero vertical deflection with the bridge balanced.

4. Hold the camera shutter open, stress the material or structure under test, close the camera shutter, and develop the picture.

5. Using the developed photo, measure the strain versus time plot. Use a longer sweep time interval if the complete strain plot is not displayed.

### 12-17.2. Acceleration Measurements

The procedure for acceleration measurement is almost identical to that of strain measurement. The major difference is that a resistance-type accelerometer is used as the transducer in place of the strain gauge. The opposite leg of the bridge is composed of a fixed resistance to match the accelerometer resistance. The bridge is balanced under no-acceleration conditions or at some preselected value of acceleration by potentiometer $R_1$. The oscilloscope is not deflected when the bridge is balanced (no acceleration). When the acceleration changes, the accelerometer resistance is changed. This unbalances the bridge and produces a d-c output that is proportional to *acceleration change*. The d-c output deflects the oscilloscope vertical trace and produces a plot of acceleration change versus time. The bridge can be calibrated in terms of acceleration (feet per second$^2$ per volt) or whatever is convenient. The entire trace can be photographed for a permanent record. (Set up the equipment as shown in Fig. 12-32 and follow the procedure of Section 12-17.1.)

### 12-17.3. Pressure Measurements

The procedure for pressure measurement is almost identical to that for strain measurement. The major difference is that a resistance-type pressure transducer is used in place of the strain gauge. These pressure transducers or gauges are actuated by a bellows or diaphragm that moves with pressure changes. Bellows or diaphragm movement causes a corresponding change in resistance. The opposite leg of the bridge is composed of a fixed resistance to match the pressure transducer resistance. The bridge is balanced under no-pressure conditions or at some preselected value of pressure by potentiometer $R_1$. The oscilloscope is not deflected when the bridge is balanced (no pressure).

When pressure changes, the pressure transducer resistance changes.

This unbalances the bridge and produces a d-c output that is proportional to pressure change. The d-c output deflects the oscilloscope vertical trace and produces a plot of *pressure change versus time.* The bridge can be calibrated in terms of pressure (pounds per volt) or whatever is convenient.

Pressure is sometimes measured with piezoelectric pressure pickups as shown in Fig. 12-32. These pickups are self-generating and do not require a bridge circuit. Piezoelectric pickups are connected directly to the vertical input of the oscilloscope. (Set up the equipment as shown in Fig. 12-32 and follow the procedure of Section 12-17.1.)

## 12-18. Vibration Measurements

The procedure for measuring vibration (continuous, random, or impact) with an oscilloscope is almost identical to that for measuring noise (Section 12-14). The major difference is that a vibration pickup is used instead of a microphone. The pickup acts as a transducer and converts vibration into an electrical signal. Most vibration transducers are piezoelectric instruments that produce an a-c voltage proportional to the acceleration of the vibrating body. Some vibration transducers contain integration networks that provide output voltages proportional to velocity and displacement. Other vibration transducers are supplied with special preamplifiers. The vibration transducer manufacturers often supply calibration data that permit the voltage readings on the oscilloscope to be converted to a direct readout of velocity, acceleration, displacement, amplitude, etc. (Set up the equipment as shown in Fig. 12-33 and follow the procedure of Section 12-14.)

## 12-19. Speed Measurements

There are two basic types of speed measurement: stroboscopic and transducer. The speed of phonograph turntables, reciprocating engines, motors, and machinery with vibrating or rotating parts may be determined by either method. The strobe method does not require physical contact with the object being measured. Most transducers do require some physical contact.

There are three basic types of transducers: magnetic, capacitive, and photoelectric. The magnetic tachometer-type transducer is the most common. Such magnetic transducers are miniature generators driven by the rotating machinery. Usually, they produce an a-c output. Some magnetic transducers produce an output voltage proportional to rotational speed and are rated in RPM per volt. Other magnetic transducers produce a signal the frequency of which is proportional to rotational speed. The

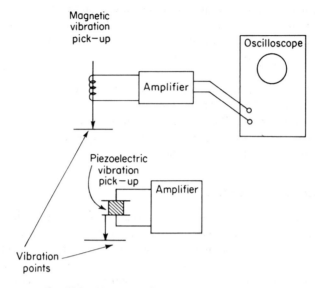

**Fig. 12-33.** Measuring vibration with an oscilloscope.

capacitive and photoelectric transducers are almost always of the frequency type.

The stroboscopic systems are based on using an audio signal to control the frequency of the ON-OFF cycle of a neon lamp.

### 12-19.1. Frequency-type Speed Measurements

1. Connect the equipment as shown in Fig. 12-34.

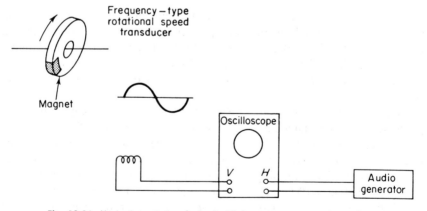

**Fig. 12-34.** Measuring rotational speed with frequency-type speed transducer.

2. Place the oscilloscope in operation as described in the instruction manual. Switch off the internal recurrent sweep. Set the sweep selector and sync selector to external.

3. Connect the transducer to the machinery as described in the transducer instruction manual. Capacitive and photoelectric transducers may or may not require direct coupling.

4. Place the audio generator in operation as described in the instruction manual.

5. Adjust the audio generator frequency to obtain a stable pattern on the oscilloscope. Identify the transducer output frequency by means of Lissajous patterns (Section 8-4).

### NOTE

In the case of a laboratory oscilloscope, the audio generator can be omitted, and frequency can be determined using the internal sweep, as described in Section 8-3.

6. Convert the frequency measurement into a speed indication, using the conversion factor supplied with the transducer.

#### 12-19.2. Voltage-type Speed Measurements

1. Connect the equipment as shown in Fig. 12-35. Note that either a meter or oscilloscope can be used as a readout with a voltage-type speed transducer.

2. If the oscilloscope is used, place it in operation as described in the instruction manual. Switch on the internal recurrent sweep. Set the sweep selector and sync selector to internal.

3. Connect the transducer to the machinery as described in the transducer instruction manual.

4. Set the meter (or oscilloscope) to measure ac or dc, depending upon the transducer output.

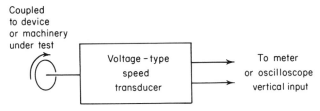

**Fig. 12-35.** Measuring speed with voltage-type speed transducer.

5. With the machinery operating, measure the transducer output voltage.

6. Convert the voltage measurement into a speed indication, using the conversion factor supplied with the transducer.

### 12-19.3. Stroboscopic Speed Measurements

#### NOTE

The following procedure shows how the stroboscopic system can be applied to measure speed of a rotating device such as a phonograph turntable. The same basic procedure can be applied to almost any rotating object.

1. Connect the equipment as shown in Fig. 12-36. Connect the positive terminal of the battery to the audio generator output.

#### NOTE

The strobe disc shown in Fig. 12-36 is made of stiff paper or cardboard and is composed of an *even number of equal sized* alternating black-and-white segments. This disc is placed on the rotating body.

2. Adjust the audio generator output for minimum.
3. Adjust the battery voltage with potentiometer $R_1$ until the neon lamp

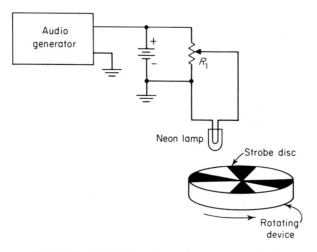

**Fig. 12-36.** Basic stroboscopic speed measurement circuit.

barely starts to glow. Then reduce the battery voltage to the point at which the lamp goes out.

4. Increase the audio generator output until the neon lamp lights.

5. Start the rotating device and hold the neon lamp close enough to the strobe disc so that the disc is illuminated. It may be necessary to darken the room or reduce the local lighting in order to obtain best results.

6. Vary the audio generator frequency, starting at the low end of the low-frequency limit of the audio generator, until the disc *appears stationary* and the disc segments are normal in width.

7. Read the audio generator frequency and calculate the speed of the rotating device using the equation

$$\text{speed (in RPM)} = \frac{60 \times \text{generator frequency in Hz}}{\text{number of black segments}}$$

For example, if the disc has eight black segments and the frequency is 700 Hz when the disc appears motionless, the speed is

$$\frac{60 \times 700}{8} = 5250 \text{ RPM}$$

### NOTE

The stroboscopic method of speed measurement can also be applied to measurement of *vibration speed,* eliminating the need for direct coupling between the vibration transducer and the object being measured. Instead, the neon lamp is held close enough to the vibrating object so that light is reflected by the object. The generator frequency is adjusted until the vibrating motion appears stopped. Vibration frequency (or speed) is then equal to the generator frequency. Note that the stroboscopic method cannot be used to measure such factors as vibration displacement, amplitude, or velocity.

# Index